Addressing Special Educational Needs and Disability in the Curriculum: Geography

The SEND Code of Practice (2015) reinforced the requirement that *all* teachers must meet the needs of *all* learners. This topical book provides practical, tried and tested strategies and resources that will support teachers in making geography lessons accessible and interesting for all pupils, including those with special needs. The author draws on a wealth of experience to share her understanding of individual needs and to show how the geography teacher can reduce or remove any barriers to learning.

Offering strategies that are specific to the context of geography teaching, this book will enable teachers to:

- ensure all pupils are able to participate in every lesson;
- provide inclusive fieldwork opportunities that take into consideration the needs of all pupils;
- use formative assessments to measure learning and inform planning;
- plan the classroom layout and display to enhance learning;
- successfully use the support of their teaching assistants.

An invaluable tool for continuing professional development, this text will be essential for middle school and secondary geography teachers (and their teaching assistants) who are committed to creating an inclusive learning environment for all pupils, regardless of their individual needs. The book will also be of interest to SENCOs and inclusion managers, curriculum co-ordinators, senior management teams and ITT providers.

Full of practical ideas and inspiration, and supported by free online resources, this is an essential tool for geography teachers and teaching assistants who want to deliver successful, inclusive lessons for all pupils.

Helen Harris is an experienced geography teacher and a consultant to the Geographical Association.

Addressing Special Educational Needs and Disability in the Curriculum

Series Editor: Linda Evans

Children and young people with a diverse range of special educational needs and disabilities (SEND) are expected to access the full curriculum. Crucially, the current professional standards make it clear that *every* teacher must take responsibility for *all* pupils in their classes. Titles in this fully revised and updated series will be essential for teachers seeking subject-specific guidance on meeting their pupils' individual needs. In line with recent curriculum changes, the new Code of Practice for SEN and other pedagogical developments, these titles provide clear, practical strategies and resources that have proved to be effective and successful in their particular subject area. Written by practitioners, they can be used by departmental teams and in 'whole-school' training sessions as professional development resources. With free web-based online resources also available to complement the books, these resources will be an asset to any teaching professional helping to develop policy and provision for learners with SEND.

The new national curriculum content will prove challenging for many learners, and teachers of children in Y5 and Y6 will also find the books a valuable resource.

Titles in this series include

Addressing Special Educational Needs and Disability in the Curriculum: Languages
John Connor

Addressing Special Educational Needs and Disability in the Curriculum: Music
Victoria Jaquiss and Diane Paterson

Addressing Special Educational Needs and Disability in the Curriculum: PE and Sport
Crispin Andrews

Addressing Special Educational Needs and Disability in the Curriculum: Science
Marion Frankland

Addressing Special Educational Needs and Disability in the Curriculum: Design and Technology
Louise T. Davies

Addressing Special Educational Needs and Disability in the Curriculum: History
Ian Luff and Richard Harris

Addressing Special Educational Needs and Disability in the Curriculum: Religious Education
Dilwyn Hunt

Addressing Special Educational Needs and Disability in the Curriculum: Art
Gill Curry and Kim Earle

Addressing Special Educational Needs and Disability in the Curriculum: English
Tim Hurst

Addressing Special Educational Needs and Disability in the Curriculum: Maths
Max Wallace

Addressing Special Educational Needs and Disability in the Curriculum: Geography
Helen Harris

For a full list of titles, please visit: www.routledge.com/series/SENCURR

Addressing Special Educational Needs and Disability in the Curriculum: Geography

Second edition

Helen Harris

Routledge
Taylor & Francis Group

LONDON AND NEW YORK

Second edition published 2018
by Routledge
2 Park Square, Milton Park, Abingdon, Oxon, OX14 4RN

and by Routledge
711 Third Avenue, New York, NY 10017

Routledge is an imprint of the Taylor & Francis Group, an informa business

First edition published by Routledge 2005

British Library Cataloguing-in-Publication Data
A catalogue record for this book is available from the British Library

Library of Congress Cataloging-in-Publication Data
Names: Harris, Helen (Geography teacher), author.
Title: Addressing special educational needs and disability in the
   curriculum. Geography / Helen Harris.
Description: 2nd edition. | Abingdon, Oxon ; New York, NY : Routledge,
   2018. | Includes bibliographical references and index.
Identifiers: LCCN 2017047043 (print) | LCCN 2018000502 (ebook) |
   ISBN 9781315457772 (ebook) | ISBN 9781138209091 (hardback) |
   ISBN 9781138209107 (pbk.)
Subjects: LCSH: Geography—Study and teaching (Elementary)—Great
   Britain. | Special education—Great Britain.
Classification: LCC G76.5.G7 (ebook) | LCC G76.5.G7 H37 2018 (print) |
   DDC 372.89/1049—dc23
LC record available at https://lccn.loc.gov/2017047043

ISBN: 978-1-138-20909-1 (hbk)
ISBN: 978-1-138-20910-7 (pbk)
ISBN: 978-1-315-45777-2 (ebk)

Typeset in Helvetica
by Apex CoVantage, LLC

 Visit the eResources: www.routledge.com/9781138209107

This book is dedicated to Patrick Eavis, retired Headmaster of Queen Elizabeth High School, Hexham, Northumberland, for his guidance and inspiration throughout my career.

Contents

Website materials

CPD audit

Model departmental policy

Example of an annotated poster 'Earthquakes'

Assessment and feedback: 'Earthquakes'

Rainforest: pre/post topic assessment

Settlements: pre/post topic assessment

TA focus – planning sheet

Series authors

The author

Helen Harris has 27 years of teaching experience across the age and ability range in three secondary schools, where she has also held a range of middle and senior management positions. She is a geography consultant to the Geographical Association and has exam board experience as a specification developer, examiner and trainer. She has been involved in initial teacher training at the University of Newcastle and has recently published a range of educational books in support of the new A Level Geography courses. She holds a BSc in Geography, a PGCE in Secondary Geography and an MA in Educational Leadership and Management.

Series editor

Linda Evans was commissioning editor for the original books in this series and has co-ordinated the updating process for these new editions. She has taught children of all ages over the years, and her posts have included those of SENCO, Local Authority (LA) Adviser, Office for Standards in Education (OFSTED) inspector and Higher Education (HE) tutor/lecturer. She was awarded a PhD in 2000 following research on improving educational outcomes for children (primary and secondary).

Since then, Linda has been commissioning editor for David Fulton Publishing (SEN) as well as editor of a number of educational journals and newsletters; she has also written books, practical classroom resources, Masters course materials and school improvement guidance. She maintains her contact with school practitioners through her work as a part-time Initial Teacher Training (ITT) tutor and educational consultant.

SEND specialist

Sue Briggs has been supporting the education and inclusion of children with special educational needs and disabilities, and their parents, for over 20 years; variously as a teacher, OFSTED inspector, specialist member of the SEN and

Disability Tribunal, school improvement partner, consultant and adviser. She holds a Masters degree in education, a first class BEd and a diploma in special education (Diploma in Professional Studies in Education distinction). Sue was a national lead for the Achievement for All programme (2011–2013) and a regional adviser for the Early Support programme for the Council for Disabled Children (2014–2015) and is currently an independent education and leadership consultant.

Sue is the author of several specialist books and publications including *Meeting SEND in Primary Classrooms* and *Meeting SEND in Secondary Classrooms* (both published by Routledge, July 2015).

Subject specialists

Art

Gill Curry was Head of Art in a secondary school in Wirral for 20 years and Advisory Teacher for Art and Gifted and Talented Strand Co-ordinator. She has an MA in Print from the University of Chester and an MA in Women's Studies from the University of Liverpool.

She is a practising artist specialising in print and exhibits nationally and internationally, running courses regularly in schools and print studios.

Kim Earle is Vice Principal at Birkenhead High School Academy for Girls on the Wirral. She has previously been a head of art and head of creative arts, securing Artsmark Gold in all the establishments in which she has worked. Kim was also formerly Able Pupils and Arts Consultant in St Helens, working across special schools and mainstream schools with teaching and support staff on Art policy and practice. She still teaches Art in a mixed ability setting in her current school and works closely with local schools and outside organisations to address barriers to learning.

Design and technology

Louise T. Davies is the founder of the Food Teachers Centre, offering advice and guidance to the Department for Education (DfE) and other organisations based on her years of experience as a teacher and teacher trainer, and her role in curriculum development at the Qualifications and Curriculum Authority (QCA) and the Royal College of Art. She led innovation at the Design & Technology Association, providing expertise for a range of curriculum and Continuing Professional Development (CPD) programmes and specialist advice on teaching standards and best practice, including meeting special educational needs. Most recently, she has worked as Lead Consultant for the

School Food Champions programme (2013–2016) and as an adviser to the DfE on the new GCSE Food Preparation and Nutrition.

English

Tim Hurst began his career as an English teacher at the Willian School in Hertfordshire, becoming Second in English before deciding that his future lay in SEND. He studied for an Advanced Diploma in Special Educational Needs and has been a SEN co-ordinator in five schools in Hertfordshire, Essex and Suffolk. Tim has always been committed to the concept of inclusion and is particularly interested in reading development, which he passionately believes in as a whole-school responsibility.

History

Ian Luff taught in comprehensive schools for 32 years and was Head of History in four such schools, writing extensively and delivering training in teaching the subject. He served in the London Borough of Barking and Dagenham as Advisory Teacher and as Deputy Headteacher at Kesgrave High School in Suffolk. Ian was made an honorary fellow of the Historical Association for contributions to education in 2011 and is currently an associate tutor and PhD student in the School of Education and Lifelong Learning at the University of East Anglia.

Richard Harris taught in comprehensive schools for 16 years and was Head of History and Head of Humanities, as well as a teacher consultant for history in West Berkshire. He has spent 15 years working with trainee history teachers at the Universities of Southampton and Reading and is currently Director of Teaching and Learning as well as researching issues mainly relating to history education. He has advised government bodies and worked extensively with the Council of Europe on teacher education and history education. He was made an honorary fellow of the Historical Association in 2011.

Maths

Max Wallace has nine years' experience of teaching children with special educational needs. He currently works as an advanced skills teacher at an inclusive mainstream secondary school. Appointed as a specialist leader in education for Mathematics, Max mentors and coaches teachers in a wide network of schools. He has previously worked as a head of year and was responsible for the continuing professional development of colleagues. He has a doctorate in Mathematics from Cardiff University.

Modern foreign languages

John Connor is a former head of faculty, local authority adviser and senior examiner. He has also served as an OFSTED team inspector for modern languages and special educational needs in mainstream settings. John was also an assessor on the Advanced Skills Teacher programme for the DfE. He is currently working as a trainer, author and consultant, and has directed teaching and learning quality audits across England, the Channel Islands, Europe, the Middle East and the Far East. He is also a governor of a local primary school.

Music

Victoria Jaquiss (FRSA) trained as a teacher of English and Drama and held the posts of English teacher, Head of PSE, Music and Expressive Arts at Foxwood School. She became a recognised authority on behaviour management and inclusion with children in challenging circumstances. The second half of her career has involved working for the Leeds Music Service/Leeds ArtForms as Steel Pan Development Officer and Deputy Inclusion Manager/Teacher. She was awarded the fellowship of the Royal Society of Arts in 2002.

Diane Paterson began teaching as a mainstream secondary music teacher. She went on to study how music technology could enable people with severe physical difficulties to make their own music, joining the Drake Music project in Yorkshire and becoming its regional leader. She then became Inclusion Manager/Teacher at Leeds Music Service/ArtForms, working with children with additional needs. As secretary of YAMSEN: SpeciallyMusic, she now runs specialist regional workshops, music days and concerts for students with special/additional needs and their carers.

PE and sport

Crispin Andrews is a qualified teacher and sports coach, and has worked extensively in Buckinghamshire schools coaching cricket and football and developing opportunities for girls in these two sports. He is currently a sports journalist, writing extensively for a wide range of educational journals, including *Special Children* and the *Times Educational Supplement*, and other publications such as *Cricket World*.

Religious education

Dilwyn Hunt taught R.E. for 18 years before becoming an adviser first in Birmingham and then in Dudley. He currently works as an independent R.E. adviser supporting local authorities, SACREs and schools. He is also in demand across the country as a speaker on all aspects of teaching R.E., in both mainstream and special settings. He is the author of numerous popular classroom resources and books and currently serves as the executive assistant on the Association of R.E. Inspectors, Advisers and Consultants.

Science

Marion Frankland, CSciTeach, has been teaching for 16 years and was an advanced skills teacher of science. She has extensive experience of teaching science at all levels in both mainstream and special schools, and has worked as a SENCO in a special school, gaining her qualification alongside her teaching commitment.

A few words from the series editor

The original version of this book formed part of the 'Meeting SEN in the Curriculum' series which was published ten years ago to much acclaim. The series won a BERA (British Educational Resources Award) and has been widely used by ITT providers, their students and trainees, curriculum and SEN advisers, department heads and teachers of all levels of experience. It has proven to be highly successful in helping to develop policy and provision for learners with special educational needs and disabilities.

The series was born out of an understanding that practitioners want information and guidance about improving teaching and learning that is **relevant to them** – rooted in their particular subject, and applicable to pupils they encounter. These books exactly fulfil that function.

Those original books have stood the test of time in many ways – their tried and tested, practical strategies are as relevant and effective as ever. Legislation and national guidance has moved on, however, as have resources and technology; new terminology accompanies all of these changes. For example, we have changed the series title to incorporate the acronym 'SEND' (special educational needs and disability), which has been adopted in official documents and in many schools in response to recent legislation and the revised Code of Practice. The important point to make is that our authors have addressed the needs of pupils with a wide range of special or 'additional' needs; some will have education, health and care (EHC) plans, which have replaced 'statements', but most will not. Some will have identified 'syndromes' or 'conditions', but many will simply be termed 'low attainers'; pupils who, for whatever reason, do not easily make progress.

This second edition encompasses recent developments in education, and specifically in the teaching of Geography. At the time of publication, education is still very much in an era of change; our national curriculum, monitoring and assessment systems are all newly fashioned, and many schools are still adjusting to changes and developing their own ways forward. The ideas and guidance contained in this book, however, transcend the fluctuations of national politics and policy and provide a framework for ensuring that pupils with SEND can 'enjoy and achieve' in their Geography lessons.

NB: The term 'parent' is used throughout and is intended to cover any adult who is a child's main care-giver.

Linda D. Evans

Acknowledgements

As with the original version of this book, written by Diane Swift, members of the Geographical Association have provided a wealth of help and guidance in putting together this new edition. Particular thanks must go to administrator and projects officer Julie Beattie, and to Stephen Schwab, Consultant to the GA and Co-Chair of the Geographical Association Secondary Phase Committee.

Thanks to James Benefield of Hodder Education for permission to use the stepped assessment based on an activity in the *Maths in Geography* book by Helen Harris.

The publishers would also like to thank:

David Evans of Fox Lane Photography

Staff and pupils of St John's CoE Middle School in Bromsgrove and Queensbury School in Erdington for allowing us to use their photographs.

Introduction

Ours to teach

Your class: 30 individuals to teach – to encourage, motivate and inspire: 30 individuals who must be seen to make good progress regardless of their various abilities, backgrounds, interests and personalities. This is what makes teaching so interesting!

Jason demonstrates very little interest in school. He rarely completes homework and frequently turns up without a pen. He finds it hard to listen when you're talking and is likely to start his own conversation with a classmate. His work is untidy and mostly incomplete. It's difficult to find evidence of his progress this year.

Zoe tries very hard in lessons but is slow to understand explanations and has difficulty in expressing herself. She has been assessed as having poor communication skills, but there is no additional resourcing for her.

Ethan is on the autistic spectrum and finds it difficult to relate to other people, to work in a group and to understand social norms. He has an education, health and care plan which provides for some teaching assistant (TA) support, but this is not timetabled for all lessons.

Do you recognise these youngsters? Our school population is now more diverse than ever before, with pupils of very different abilities, aptitudes and interests, from a wide range of cultures, making up our mainstream and special school classes. Many of these learners will experience difficulties of some sort at school, especially when they are faced with higher academic expectations at the end of Key Stage 2 and into Key Stages 3–4.

Whether they have a specific special educational need like dyslexia, or are on the autistic spectrum, or for various reasons cannot conform to our behavioural expectations – **they are ours to teach**. Our lessons must ensure that each and every pupil can develop their skills and knowledge and make good progress.

How can this book help?

The information, ideas and guidance in this book will enable teachers of Geography (and their teaching assistants) to plan and deliver lessons that will meet the individual needs of learners who experience difficulties. It will be especially valuable to subject teachers because the ideas and guidance are provided within a subject context, ensuring relevance and practicability.

Teachers who cater well for pupils with special educational needs and disabilities (SEND) are likely to cater well for *all* pupils – demonstrating outstanding practice in their everyday teaching. These teachers have a keen awareness of the many factors affecting a pupil's ability to learn, not only characteristics of the individual but also aspects of the learning environment that can either help or hinder learning. This book will help practitioners to develop strategies that can be used selectively to enable each and every learner to make progress.

Professional development

Our education system is constantly changing. The national curriculum, SEND legislation, examination reform and significant change to OFSTED inspection mean that teachers need to keep up to date and be able to develop the knowledge, skills and understanding necessary to meet the needs of all the learners they teach. High quality continuing professional development (CPD) has a big part to play in this.

Faculties and subject teams planning for outstanding teaching and learning should consider how they regularly review and improve their provision by:

- auditing

   a) the skills and expertise of current staff (teachers and assistants)

   b) their professional development needs for SEND, based on the current cohorts of pupils

      [There is an audit proforma available for download on the accompanying website]

- using the information from the two audits to develop a CPD programme (using internal staff, colleagues from nearby schools and/or consultants to deliver bespoke training)
- enabling teachers to observe each other, teach together and visit other classrooms and other schools
- encouraging staff to reflect on their practice and feel comfortable in sharing both the positive and the negative experiences
- establishing an ethos that values everyone's expertise (including pupils and parents who might be able to contribute to training sessions)
- using online resources that are readily available to support workforce development (e.g. www.nasen.org.uk/professional-learning)
- encouraging staff to access (and disseminate) further study and high quality professional development

This book, and the others in the series, will be invaluable in contributing to whole-school CPD on meeting special educational needs, and in facilitating subject-specific staff development within departments.

1 Meeting special educational needs

Your responsibility

New legislation and national guidance in 2014 changed the landscape of educational provision for pupils with any sort of 'additional' or 'special' needs. The vast majority of learners, including those with 'moderate' or 'mild' learning difficulties, weak communication skills, dyslexia or social/behavioural needs, rarely attract additional resources: they are very much accepted as part of the 'mainstream mix'. Pupils with more significant special educational needs and/ or disabilities (SEND) may have an education, health and care (EHC) plan: this outlines how particular needs will be met, often involving professionals from different disciplines, and sometimes specifying adult support in the classroom. Both groups of pupils are ultimately the responsibility of the class teacher, whether in mainstream or special education.

> High quality teaching that is differentiated and personalised will meet the individual needs of the majority of children and young people. Some children and young people need educational provision that is additional to or different from this. This is special educational provision under Section 21 of the Children and Families Act 2014. Schools and colleges **must** use their best endeavours to ensure that such provision is made for those who need it. Special educational provision is underpinned by high quality teaching and is compromised by anything less.
>
> SEND Code of Practice 2015 (DfE, 2015)

There is more information about legislation (The Children and Families Act 2014; The Equality Act 2010) and guidance (SEND Code of Practice) in Appendix 1.1.

Definition of SEND

A pupil has special educational needs if he or she:

- has a significantly greater difficulty in learning than the majority of others of the same age; or
- has a disability which prevents or hinders him or her from making use of facilities of a kind generally provided for others of the same age in mainstream schools or mainstream Post-16 institutions.

(SEND Code of Practice 2015)

The SEND Code of Practice identifies four broad areas of SEND (Table 1.1), but remember that this gives only an overview of the range of needs that should be planned for by schools; pupils' needs rarely fit neatly into one area of need.

Whole-school ethos

Successful schools are proactive in identifying and addressing pupils' special needs, focusing on adapting the educational context and environment rather than on 'fixing' an individual learner. Adapting systems and teaching programmes rather than trying to force the pupil to conform to rigid expectations will lead to a greater chance of success in terms of learning outcomes. Guidance on whole-school and departmental policy making can be found in Appendix 1.2, and a sample departmental policy for SEND can be downloaded from our website.

Table 1.1 The four broad areas of SEND

Communication and interaction	Cognition and learning	Social, emotional and mental health difficulties	Sensory and/or physical needs
Speech, language and communication needs (SLCN)	Specific learning difficulties (SpLD)	Mental health difficulties such as anxiety or depression, self-harming, substance abuse or eating disorders	Vision impairment (VI)
Asperger's syndrome and autism (autistic spectrum disorder (ASD))	Moderate learning difficulties (MLD)		Hearing impairment (HI)
	Severe learning difficulties (SLD)		Multi-sensory impairment (MSI)
	Profound and multiple learning difficulties (PMLD)	Attention deficit disorders, attention deficit hyperactivity disorder or attachment disorder	Physical disability (PD)

Policy into practice

In many cases, pupils' individual learning needs will be met through differentiation of tasks and materials in their lessons; sometimes this will be supplemented by targeted interventions such as literacy 'catch-up' programmes delivered outside the classroom. A smaller number of pupils may need access to more specialist equipment and approaches, perhaps based on advice and support from external specialists.

The main thrust of the Children and Families Act and Chapter 6 of the SEND Code of Practice is that outcomes for pupils with SEN must be improved and that schools and individual teachers must have high aspirations and expectations for all.

In practice, this means that pupils should be enabled to

- *achieve their best:* additional provision made for pupils with SEND will enable them to make accelerated progress so that the gap in progress and attainment between them and other pupils is reduced. Being identified with SEND should no longer be a reason for a pupil making less than good progress.
- *become confident individuals living fulfilling lives:* if you ask parents of children with SEN what is important to them for their child's future, they often answer 'happiness, the opportunity to achieve his/her potential, friendships and a loving family' – just what we all want for our children. Outcomes in terms of well-being, social skills and growing independence are equally as important as academic outcomes for children and young people with SEND.
- *make a successful transition into adulthood, whether into employment, further or higher education or training:* decisions made at transition from primary school, in Year 7 and beyond should be made in the context of preparation for adulthood. For example, where a pupil has had full-time support from a teaching assistant in primary school, the secondary school's first reaction might be to continue this level of support after transition. This may result in long-term dependency on adults, however, or limited opportunities to develop social skills, both of which impact negatively on preparation for adulthood.

Excellent classroom provision

Later chapters provide lots of subject-specific ideas and guidance on strategies to support pupils with SEND. In Chapter 3 you will find useful checklists to help you support pupils with identified 'conditions', but there are some generic approaches that form the foundations of outstanding provision, such as:

- providing support from adults or other pupils;
- adapting tasks or environments;
- using specialist aids and equipment as appropriate.

The starting points in the following list provide a sound basis for creating an inclusive learning environment that will benefit *all* pupils, while being especially important for those with SEND.

Develop pupils' understanding through the use of all available senses by

- using resources that pupils can access through sight *and* sound (and where appropriate also using the senses of touch, taste and smell to broaden understanding and ensure stronger memory); see Appendix 1.3;
- regularly employing resources such as symbols, pictures and film to increase pupils' knowledge of the wider world and contextualise new information and skills;
- encouraging and enabling pupils to take part in activities such as drama, class visits, field trips and exploration of the environment.

Help pupils to learn effectively and prepare for further or higher education, work or training by

- setting realistic demands within high expectations;
- using positive strategies to manage behaviour;
- giving pupils opportunities and encouragement to develop the skills to work effectively in a group or with a partner;
- teaching all pupils to value and respect the contribution of others;
- encouraging independent working skills;
- teaching essential safety rules.

Help pupils to develop communication skills, language and literacy by

- making sure all pupils can see your face when you are speaking;
- giving clear, step-by-step instructions and limiting the amount of information given at one time;
- providing a list of key vocabulary for each lesson;
- choosing texts that pupils can read and understand;
- making texts available in different formats, including large text or symbols, or by using screen reader programs;
- putting headings and important points in bold or highlighting to make them easier to scan;
- presenting written information as concisely as possible, using bullet points, images or diagrams.

Support pupils with disabilities by

- encouraging pupils to be as independent as possible;
- enabling them to work with other, non-disabled pupils;
- making sure the classroom environment is suitable, e.g. uncluttered space to facilitate movement around the classroom, and adapted resources are labelled and accessible;

- being aware that some pupils will take longer to complete tasks, including homework;
- taking into account the higher levels of concentration and physical exertion required by some pupils (even in activities such as reading and writing) that will lead to increased fatigue for pupils who may already have reduced stamina;
- being aware of the extra effort required by some pupils to follow oral work, whether through use of residual hearing, lip reading or signed support and of the tiredness and limited concentration which is likely to ensue;
- ensuring all pupils are included, and can participate safely, in fieldwork, school trips and off-site visits.

These and other, more specific strategies are placed in the context of supporting particular individuals in Chapter 3 and Appendix 3.1.

2 Creating the inclusive classroom

What is the classroom environment?

The classroom environment is defined as the physical and emotional context of an educational setting, in other words, how it looks and feels. The classroom environment is made up of a set of interacting factors involving physical, social and emotional considerations.

All schools develop their physical site for maximum inclusion and should work with individual teachers to develop classrooms in the same way.

Why is the classroom environment important?

The classroom can be either a powerful motivational aid to teaching or an uninviting and disorganised negative influence. Classrooms should be inviting environments that make pupils feel good about being there; it is well documented that pupils learn best when they are at ease, relaxed and confident enough to ask for help. From the moment of entering, the classroom environment can invoke positive or negative feelings directly or indirectly for each pupil; these feelings may influence behaviour in the early stages of a lesson, or may form a more lasting impression. Teachers can influence learning, achievement and behaviour by creating a safe, stimulating and inviting environment for all pupils and particularly for those with special educational needs (SEN). Setting the learning context represents an important component of classroom management.

There are a number of research studies demonstrating the importance of 'environment' for learning; this includes a UK study which provides strong evidence for the impact of specific aspects of classroom design (lighting, temperature and colour) on student achievement (Barrett et al., 2015).

To include pupils in geography lessons, the following general principles should be followed:

- All pupils have equal opportunity to experience and participate in the lesson.
- Appropriate levels of support are provided to aid independent learning.
- The physical environment of the classroom allows freedom of movement and access to activities and resources.

Additionally, for teachers who have a classroom which is not shared with others, it can become a reflection of your personality and teaching approach. If you are comfortable and happy in your teaching space, then this will enhance your own well-being and personal motivation, and ultimately your effectiveness with your pupils.

Different aspects of the classroom environment

The physical environment

Within the classroom inevitably there are places where there is a better view, clearer sound, better lighting, less distraction, easier access. Teachers must plan the most productive and supportive working environment for all students in the class and pay particular attention to individuals with specific needs. Regarding the physical environment, consideration should be given to access for wheelchair users, careful positioning for pupils with hearing or visual impairment, an understanding that pupils with attention deficit hyperactivity disorder (ADHD) and dyslexia are better away from the distraction of windows and open doors and ensuring that all pupils have enough room where seated to maximise their ability to focus. Pupils on the autistic spectrum may benefit from a carrel or being seated in a partially screened area of the room. The following covers the main considerations of the physical classroom environment.

Seating

Seating layouts are of particular importance. Appropriate seating can aid a learning task, solve behavioural issues and accommodate special needs. Pupils who are visually impaired need to be near the main point of communication from the teacher, whether he or she is writing on a whiteboard, using an interactive whiteboard or simply speaking (important for lip reading). This is an important point as teachers often like to walk around the room whilst talking and inadvertently make lip reading and reading of facial expression difficult.

Pupils with hearing difficulties need to be seated where there is clarity of sound, i.e. away from windows and doors with external noise. Pupils with ASD can have a problem with eye contact, and so seating them directly in front of you may make them feel uncomfortable; a location where it would be

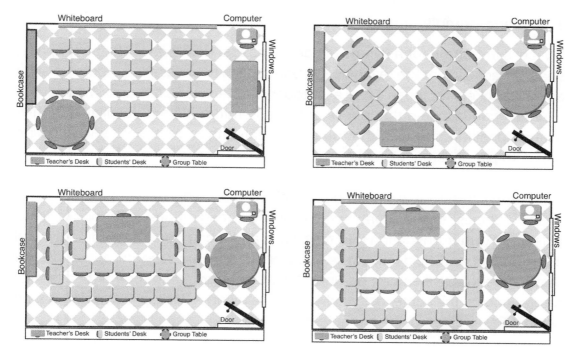

Figure 2.1 Alternative classroom layouts

easy to access a quiet seating area of the classroom for some 'time out' can help if the room is large enough. Pupils with ADHD will benefit from a position which helps them to focus on the teacher, and the needs of dyslexic pupils are best met through a position with a clear view of the board, near the teacher and with consideration to the amount of space needed, i.e. perhaps a double desk – geography lessons often draw upon a range of resources (a textbook, handouts of maps and diagrams, worksheets), and a dyslexic pupil may have difficulty organising everything in a small space. Left-handed pupils will benefit from having plenty of space on their left side. Figure 2.1 shows a selection of the many variations of classroom layout.

The correct seating arrangement can have a significant impact on pupils' learning and on classroom management. There is always a balance to be achieved between allowing pupils to sit where they choose and engaging in 'intervention', such as splitting up disruptive pupils, not allowing attention seekers to take centre stage and involving quieter pupils in terms of where they sit. Placing students requires sensitivity and discretion. In all instances of special seating arrangements, it is advisable to consult with the Special Education Needs and Disability Co-ordinator (SENDCO) and make sure that arrangements are agreed prior to the lesson so that pupils can enter the classroom and be seated without direction. Never draw attention to a pupil in front of the whole class or make seating changes once the class is settled unless it is a behaviour issue which requires immediate action.

Space and organisation

Space is important for a variety of special needs: desk space, mobility space, space to accommodate teaching support, space to access requirements such as plug sockets for those using a laptop for written work. It is also important to maintain continuity so that pupils become comfortable and familiar with their seating position and with the routine for entering a classroom and knowing where to place personal items.

Geography classrooms should be organised in a logical fashion; there may be specific 'zones' of the room, such as a small reading area with books, journals and Further Education (FE) resources and a computer area for individual investigation and fieldwork write-up. Zoning different uses of space means that different activities can be undertaken as appropriate; not all pupils may be working on the same task at all times. Fieldwork write-up is certainly a time when a flexible approach is required.

Storage areas for equipment and geographical resources should be clearly marked, accessible to students and kept tidy. Figure 2.2 illustrates some basic resources for all teachers to have to hand in their geography classroom, and the section on IT resources on p. 18 points out some specific technological resources.

Lighting

As much natural light as possible is most effective, but teachers will have to work with what the location of the classroom allows. Electric lighting strips,

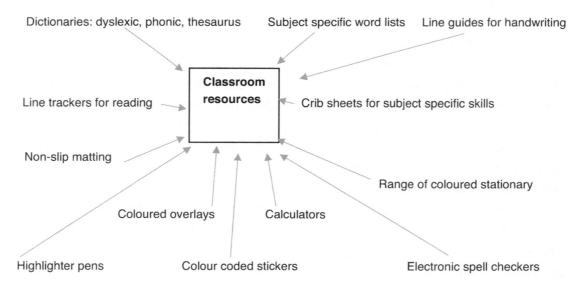

Figure 2.2 Basic classroom resources

particularly if fluorescent lights flicker, can be a real issue for most pupils, especially those with autism, anxiety and/or sensory processing disorders. Use blinds to reduce sunlight glare where pupils are working. Higher quality electrical lighting is beneficial, as is the use of lights with non-fluorescent bulbs. A work station with additional lighting, e.g. a desk lamp, may be needed by any student with visual impairment.

Acoustics

A good acoustic environment is beneficial to all students, but especially those with imperfect hearing. It's important to remember that hearing/radio aids amplify all sounds – those we want to hear and others that we don't. Classrooms can be improved acoustically by having soft furnishings such as carpets and curtains/blinds, as well as rubber ends on chair legs. Minimise external noise by closing doors and windows if necessary and keeping general classroom noise to a minimum.

Air quality

Temperature is also an important environmental factor; a thermostat for the room is most effective, but if this is not possible, fans, blinds and windows help. Classrooms should also have appropriate ventilation to allow for good airflow and quality.

Decoration

Walls should be a light neutral colour such as white or beige. Vivid colours can be used to draw the eye to important things, but displays and decorations should co-ordinate with one another and should be pleasing to the eye and not garish or confusing. Too many brightly coloured posters or decorations are not supportive of student learning.

Displays

Displays are a key addition to classroom 'decoration' and can provide a useful learning and motivational resource. It is worth keeping some general points in mind.

- Word walls display key geographical terminology for current topics being studied. Using the correct terminology receives credit in exams and is extremely important in geography. Many students struggle with new vocabulary, and word walls can create a useful reference point and an element of passive learning. (See Appendix 2.1 for an example of key words for Tourism – for display or use as a 'word mat'.)

- Sequencing of explanations is a key skill in geography. A clear display can be used to remind pupils of sequencing structures. These can be focused on the current topic under study, be it physical, human or environmental geography. Some examples of sequencing of information are shown in Figure 5.1 on p. 80.
- At exam time, it can be useful to remind students of the meaning of command terms in exam questions through a display or an annotated question.
- All GCSE and A Level geography specifications require fieldwork and at A Level the production of an individual study. Such tasks draw on a wide range of skills that students with special needs will find challenging – organisation of work into chapters, data presentation, analysis of data, photographs and sketches. Displays can be used at such times to provide a constant reference point, e.g. to show the different sections/chapters of fieldwork or individual study in the correct order, to display methods of data presentation or to help with the analysis of photographs as geographical evidence.
- Displays showing pupils' work can recognise achievement and increase motivation.
- Always take care to not 'over-stimulate' with the amount of displays and information. Perception of the world around them can be confusing for some pupils with SEN, so overloading the senses is not good, particularly for those with conditions such as ADHD or those on the autistic spectrum.

Boards

There may be different boards used for different purposes, e.g. for the lesson objective, which can remain in place; for homework reminders; for key words used in that lesson. Think about the clarity of writing on boards – size, spacing, colour, use of bullet points, clarity of diagrams. Do not produce large amounts of text on boards, which may be confusing and difficult to read, especially for any student sitting facing in the opposite direction. Written transcripts should be handed out to reduce tasks such as copying from the board. Allow plenty of time for pupils to read and respond to what is on the board. In terms of the physical setting, think about a central placing of the whiteboard with a clear view with no glare or reflection, and as a general point consider that an off-white, cream or pale pastel colour is a better background colour for students with special learning needs such as dyslexia. Black on white is often the least easy-on-the-eye combination.

The specific use of interactive whiteboards in geography is covered in the section 'Using Information Technology in Geography Lessons'.

Classroom furniture

The provision of good quality desks, chairs, flooring and other furnishings has safety implications, e.g. regarding stability if pupils need to raise and lower themselves into seating or raise a table top to accommodate a wheelchair. Good quality, attractive furniture can also promote a sense of student pride and well-being, as well as ensuring a degree of comfort that will maximise concentration and effective learning.

The teacher

There is the important issue of where to position yourself as teacher. It goes without saying that teachers should move around the classroom, engage with all pupils and position themselves where they are in clear view. They should project themselves with clarity, give visual cues if helpful and be mindful of tone and of the possibility of being taken literally. The teacher's desk is also worth some thought. What is it for, and who needs to access it? If it is used to contain resources and locate a computer for registration and whiteboard use, then presumably it can be quite small and in the line of sight but does not need to present a bold statement blocking the view of the board.

The social environment

The social environment concerns how the teacher connects with their pupils and how the pupils socialise with each other. It is essential that teachers become familiar with who their pupils are in order to manage the social interaction, e.g. making some use of peer support. However, be mindful not to overburden a pupil by expecting them to fulfil the role of a support assistant.

Managing the social environment is an ongoing process, but in the initial stages of planning an inclusive classroom, a teacher must carry out a degree of preparation through consultation with SENDCOs, tutors, pastoral heads and support teachers and through some research on how to support pupils with specific needs. Teachers need to address two key questions in relation to the social environment of the classroom:

- What are the needs of SEN pupils in my classes?
- How can their learning be improved through their social interactions?

The emotional environment

There is a clear research consensus which connects pupils' emotional needs to their ability to focus on learning. When a pupil's needs for belonging, well-being and self-esteem are met, an important foundation for learning is established. Researchers have for a long time identified specific connections between classroom environments and pupils' academic success.

Work by Osterman (2000), in educational psychology, offered further support to the principle that when the emotional and psychological needs of individuals are met, particularly the need for a sense of inclusion in a supportive environment, then there is an impact on behaviour. This connection, where a pupil's emotional well-being impacts on motivation, engagement, learning and achievement, can operate within both the whole-school environment and the individual classroom.

> Children are not going to engage in their learning if they're not interested, if their needs aren't being met, if they're not supported, if they're experiencing difficulties, or if they don't think that there's anything in it for them.
>
> C Norman, Welbeck Primary School

The emotional needs of pupils include the following.

• The need for respect

Within the classroom there is a fundamental need for mutual respect. Respect is shown by listening to others when they speak and volunteer answers, sensitively addressing misconceptions and wrong answers, and fostering an appreciation that everyone has a right to learn and that it is wrong to jeopardise that right through bad or insensitive behaviour. The teacher may instil respect by making the point that calling out is not acceptable or that it is sometimes necessary to wait for help if the teacher is not readily available (though a quick acknowledgement from the teacher may reassure a pupil and encourage patience).

• The need to enhance motivation and self-esteem

There is nothing more motivational and uplifting to any pupil than praise. Praise and encouragement are the building blocks of emotional and social development; however, it is worth noting several points:

• Consistency is everything.
• Praise should be appropriately earned, i.e. given for 'real' effort and achievement, so that it is not devalued.
• Praise may be offered in a variety of forms – written or a 'quiet' word.
• Actions can also be used, such as an encouraging smile or some 'one to one' attention.

Remember that praise need not relate only to academic achievement; it can also be given for good behaviour, supportive pair work, a kind gesture or to recognise and reinforce the extra effort made by a pupil with special needs.

Teachers need to know that in their classes, not all pupils like to receive public praise. Praise should never be used for appeasement and is devalued by overuse. Indeed, some research also suggests that 'lavish' praise does little to help pupils and that what is essential is that 'teachers need to understand why, when and how a particular approach is likely to enhance a student's learning' (Coe et al., 2014).

- The need for sensitivity

There is a wide range of circumstances requiring sensitivity. Teachers need to be aware of the fact that what may seem a 'throwaway comment' to them can be interpreted very differently by an individual or a group of teenagers. Pupils with certain conditions such as autism also take things very literally. Knowing how to use pupils' strengths and not expose points of weakness is an important teaching necessity based on awareness, empathy and sensitivity. Situations to be aware of include physical impediments and a lack of confidence in a certain area, e.g. reading aloud, making verbal contributions or sharing thoughts.

Geography as a subject covers controversial issues often requiring sensitive handling. Topics relating to race, poverty, development, culture and government policy will require forethought in handling. An individual's ethnicity or personal circumstances may warrant a degree of sensitive handling by the teacher and from other pupils.

- The need to foster a sense of belonging

It is not just desirable but essential that the 'class' works as a cohesive team unit and that all individuals are respected and valued as part of this. Pupils with SEND need to be fully included as 'team members' and not viewed in isolation, particularly if working with a teaching assistant (TA). Always make time to monitor their progress within the lesson, particularly if a TA is supporting; do not leave it to a conversation at the end of the lesson. Include pupils with SEND in class discussions if they are keen to participate, showing patience and offering encouragement. Make sure that they are working towards the same learning objectives (and behavioural boundaries) as the rest of the class and that differentiated tasks and resources are distributed sensitively and with discretion, perhaps having handed them to a TA prior to the lesson.

- The need for confidence building

A confident learner is a successful learner. Confidence must be nurtured at the whole-class and the individual level. A pupil with special needs will feel an additional sense of belonging and personal well-being if working within a class that has a positive outlook and a confident approach to learning. This confidence is established over time as a teacher engages with the class; builds rapport; provides interesting, stimulating and challenging learning activities; and takes account of individual needs in their lesson planning.

Resourcing the geography classroom

Using worksheets in geography lessons

- Worksheets are frequently used in geography to support explanations, for example, of how landforms are produced in physical geography or of concepts in human geography. For all worksheets, consider the language, terminology, font, size of print, highlighting, breaking down of chunks of text, clarity, spacing and colour.
- For many students with special needs, cream is often a better background colour than white.
- Line guides, highlighter pens and accessible dictionaries also help students.
- Worksheets can easily be differentiated by the amount of text and the level of completion required by the student.
- Outline maps (maps are given separate consideration in the section 'Using Maps in Geography'), data, graphs, diagrams, photographs and sketches are used in geography to provide evidence and illustrate concepts. Again consideration should be given to clarity, clear labelling or annotation, and appropriate text in terms of size, style, reading age and quantity.
- In addition to the content and quality of worksheets, consideration must be given to how pupils retain and organise these resources so that they do not become a wasted resource discarded at the back of a book or file. For these to be of use, pupils need time and support to organise and locate worksheets within their file of work, and overloading with worksheets will not be desirable or helpful to any student with learning difficulties.

Using information technology in geography lessons

Information technology (IT) is the medium of today for all pupils, offering enormous quantities of information in a highly engaging and accessible format. The use of technology has widely recognised benefits in education, such as engagement, sustained concentration and personal creativity.

Technology offers enormous quantities of information in engaging and accessible formats

In the study of geography, IT resources are more than just a source of information; they are used to analyse, investigate, organise, evaluate, communicate and present information. General uses of technology in geography include written communication, research, mapping, use of geographical information systems (GIS), working with data, use of visual images, understanding processes and fieldwork. This is very broad topic area, and opportunities and learning activities exist beyond what can be covered here: there are many resources and courses available to geography teachers covering the use of technology and how it can specifically help and support pupils with SEN. The following puts forward some initial guidelines.

- There is a range of technological aids that can be used to help SEND pupils: 'C-pens' scan and read text to pupils, visualisers can help the visually impaired, and laptops are commonly in use for those pupils who have difficulty recording written work. (Regarding the latter, it is important for a teacher to review this work to check for accuracy and to help pupils with the organisation of electronically stored notes.) Clearly there is an important issue of cost where resources such as C-pens are concerned.
- Software such as the 'Clicker' programmes offer a wide variety of word-processing features such as word banks and writing frames which enable students to produce better quality work in a shorter time. You will need to

know your way around such software to make the best use of it, but this is a case where 'spending time' results in 'saving time'. Some schools have trained TAs to be competent and enthusiastic ambassadors for this sort of technology.

- There is a wealth of resources available now to be used with interactive whiteboards to aid the teaching of geography and to help with visualisation issues which are common in geography regarding scale, processes and understanding of distant places.
- Current textbooks usually come with electronic versions for interactive whiteboards and also often include differentiated resources. It is extremely helpful to have a textbook displayed on a board at the front of the room so that the teacher can highlight and enlarge text and explain diagrams. Also, pupils' focus is up and forward rather than having them looking down at a book in isolation. Using a visualiser can also be effective.
- Multimedia software can help pupils to investigate geographical topics that are very remote or difficult to access.
- Remember issues of concern with the physical teaching environment, such as the location of plug sockets for laptops, covered wires, good positioning for viewing, reduction of glare, and use of beige or cream rather than white as the background colour.
- There are many useful websites for free resources; some are listed in 'References and Further Information'.

Using maps in geography

Maps are worthy of separate consideration as they are a key visual resource used in geography. Technology has created new ways of presenting and using maps; they can now be interactive, of a variety of forms and scales and supported by other online learning activities and resources. Interactive atlases, maps and GIS are central to geography and widely available. The internet also offers many free resources such as web-based digital maps and historical maps. The Ordnance Survey provides free maps for schools, as does Digimap for Schools.

Maps are complex resources, and the amount of information contained in them can be overwhelming for students with SEN. Interactive whiteboards are helpful to isolate a section of a map and allow the teacher (and students) the opportunity to annotate and highlight the focal points. Maps are particularly difficult to interpret if a pupil is colour blind, and some carefully selected peer support or an annotated copy may be required. Further general points to make maps accessible to pupils include the following.

- Use, alongside the map, a handout with the area to be used heavily simplified and using thick lines.

Use a range of plans and maps matched to pupils' stages of development

- Reduce the information complexity; think about what the learning objectives are, why you are using the map and what specific information is needed by the student.
- Emphasise the line work on the map.
- Sonic and tactile maps may be accessible by linking up with a local university geography department.
- Think about creating digital images to go alongside regularly used maps.
- Perhaps create a sensory box relevant to particular landscapes represented on a map.
- Photographs related to specific map locations will also be of benefit.
- Ensure that text and graphics stand out against their backgrounds.

Worksheets, IT resources and maps have been given some specific attention as part of this chapter, and all are worthy of further consideration; see 'References and Further Information' for useful websites.

Literacy and numeracy

All subjects draw on skills in literacy and numeracy to some extent. Geography draws on literacy skills to provide explanations of landform development, to explain concepts and issues and to analyse and evaluate fieldwork and data.

The language of geography may be challenging for many students – for example:

- the specific geographical use of everyday words such as 'mouth of the river', 'water table';
- terms specific to geography, such as 'erosion'; and
- terms like 'climate', 'gradient', 'height' or 'distance', which can create barriers for many students because of their abstract nature.

Comparisons between places or peoples can create barriers for students with communication impairments, including students on the autism spectrum, because of the language needed to conceptualise how a place is the same as or different from somewhere else. Plan to teach new language explicitly. Give students opportunities to answer open-ended questions – e.g. "Why did the river flood?" Take care with using analogies, including, for example, the use of cartoon imagery to illustrate social issues.

'Including Students with SEN and/or Disabilities in Secondary Geography', http://dera.ioe.ac.uk/13793/1/geography.pdf.

Numeracy is an increasingly important element of GCSE and A Level examinations and is required in the completion of fieldwork. Table 2.1 highlights some potential difficulties with literacy and numeracy for pupils with SEND and outlines learning activities which may help.

Table 2.1 Overcoming difficulties with literacy and numeracy

Difficulty	Activity
Limited language to describe places	• Stepped development from words to sentences, use of adjectives to build from • Encourage the use of models to help work out descriptions from a photograph – spot colours, shapes, patterns etc. • The use of word mats may also be helpful here (see Appendix 2.1).
Poor imagination of places	• Read accounts of places from various sources. • Use multi-sensory resources.
Making distinctions between different types of writing	• Exercises to determine the difference between what is true/false, fact/opinion, theory/fact • Pupils turn sentences from fact to opinion etc.
Difficulties in recalling and applying geographical enquiry to writing structures	• Use writing frames to promote the organisation and structure of ideas (description, explanation, discussion, persuasion). See Appendices 2.2–2.4. • Use images to support writing, and include storyboards.
Abstracting meaning from graphs	• Living graphs, matching events in a story to those on the graph. • Regular practice of reading off graphs, including activities where students calculate, interpret and extrapolate

Difficulty	Activity
Understanding statistics and ratios	• Matching diagrams to numbers • Calculating • Pictograms
Ability to recall shapes and locations	• Informal place tests • Using shape of countries to help locate features • Use clear outlines. • Use appropriate colours. • Mapping from memory.
Imagining places from maps	• Link photos with maps. • Read text descriptions of places. • Use taped descriptions. • Empathise, experience virtual fieldtrips
Interpreting patterns	• Display specialist vocabulary in a variety of visual ways, for example nucleated, scattered, clustered, linear, adjacent, meandering

How can you know that the classroom environment is effective?

The more ownership teachers and students feel they have of their learning space the better. Allowing students to help design certain learning zones, for example deciding on seating arrangements, can be helpful and positive. Pupil feedback is an extremely useful way of fostering a sense of respect, inclusion and belonging, as referred to in this chapter, and yet it is feared by some teachers. A constructive feedback questionnaire could be extremely useful, and pupils will pick up on the fact that the teacher values their opinion. In addition:

- Classrooms can easily be audited and recommendations made.
- Specialist equipment can be expensive, but there are many things that teachers can do which are low cost and arise from knowing one's pupils and being aware of their needs.
- Close liaison with the school's SENCO should ensure up-to-date information about pupils and the specific strategies to use.
- Support for pupils with SEN may come from different sources – the teacher, support staff, the SENCO, external specialists, parents. The teacher will use this support and guidance to create an inclusive learning environment.

Observations and feedback by the SENCO and members of Senior Management Team (SMT) will inform your evaluation of the environment you have created in the classroom. However, the best measure of effectiveness will be the outcomes – for your pupils, supporting TAs and yourself. These can be quickly assessed in terms of pupils' involvement and enjoyment (on-task behaviour, positive attitudes, interested questions, completion of tasks) and their levels of progress and achievement.

3 Different types of SEND

How their characteristics might affect learning and teaching in geography

Introduction

This chapter offers starting points for information on the special educational needs most frequently encountered in mainstream schools. It describes the main characteristics of each area of special educational need and disability (SEND) with practical ideas for use by geography teachers and teaching assistants (TAs), and contacts for further information.

There is a measure of repetition, as some strategies prove to be effective with a whole range of pupils (and often with those who have no identified SEND!). However, the layout enables readers an 'at a glance' reminder of effective approaches and facilitates copying for colleagues and TAs.

The SEND Code of Practice (Department for Education, 2015) outlines four broad areas of need. These are:

- communication and interaction
- cognition and learning
- social, emotional and mental health difficulties
- sensory and/or physical needs

These broad areas are not exclusive, and pupils may have needs that cut across some or all of them. Equally, pupils' difficulties and needs will change over time. The terms used in this chapter are helpful when reviewing and monitoring special educational provision, but pupils' individual talents and interests are just as important as their disability or special educational need. Because of this, specific terms or labels need to be used with care in discussion with parents, pupils or other professionals. Unless a pupil has a firm diagnosis, and parents and pupil understand the implications of that diagnosis, it is more appropriate to describe the features of the special educational need rather than use the label. For example, a teacher might describe a pupil's spelling difficulties but not use the term 'dyslexic'.

There is a continuum of need within each of the special educational needs and disabilities listed here. Some pupils will be affected more than others and will show more or fewer of the characteristics described. Appendix 3.1 describes some 'real pupils in real classrooms' who have been identified as having various types of SEND.

Pupils with other, less common special educational needs may be included in some schools, and additional information on these conditions may be found in a variety of sources. These include the school SENCO, local authority support services, educational psychologists and online information, for example on the Nasen SEND Gateway and disability charity websites such as Mencap, CAF or I CAN, the Children's Communication Charity.

> www.nasen.org.uk
> www.mencap.org.uk
> www.cafamily.org.uk
> www.ican.org.uk

Different types of SEND featured in this section

Asperger's syndrome
Attention deficit disorder/attention deficit hyperactivity disorder
Autism/autistic spectrum disorder
Cerebral palsy
Down's syndrome
Foetal alcohol syndrome
Fragile X syndrome
Learning difficulty/disability
Sensory impairment
Social, emotional and mental health needs
Specific learning difficulties: dyslexia, dyscalculia, dyspraxia
Speech, language and communication needs
Tourette's syndrome

Asperger's syndrome

Asperger's (or Asperger) syndrome is a type of autism. People with Asperger's syndrome have average to high intelligence but experience difficulties associated with being on the autistic spectrum. They often want to make friends but do not understand the complex rules of social interaction. They have impaired fine and gross motor skills, with writing being a particular problem. Boys are more likely to be affected – with the ratio being 10:1 boys to girls. Because they appear 'odd' and naïve, these students are particularly vulnerable to bullying.

Main characteristics

*   Social interaction
    Students with Asperger's syndrome want friends but have not developed the strategies necessary for making and sustaining friendships. They find it very difficult to learn social norms. Social situations, such as lessons, can cause anxiety.

*   Social communication
    Students have appropriate spoken language but tend to sound formal and pedantic, using little expression and an unusual tone of voice. They have difficulty using and understanding non-verbal language, such as facial expression, gesture, body language and eye contact. They have a literal understanding of language and do not grasp implied meanings.

*   Social imagination
    Students with Asperger's syndrome need structured environments and routines they understand and can anticipate. They excel at learning facts and figures, but have difficulty understanding abstract concepts and generalising information and skills. They often have all-consuming special interests.

ASPERGER'S SYNDROME PROFILE

Potential impact of characteristics on learning	• May interpret language literally.
	• May speak fluently but have difficulties with participating in effective two-way communications, e.g.
	  • taking turns in conversation;
	  • not understanding and therefore being unable to adapt to the responses of others who are listening to them;
	  • continually talking about their own special interests and/or making comments that could be seen as offensive;
	  • appearing insensitive to the feelings of others.

- Language may be over-precise or too literal, and misunderstanding may occur over multiple levels of meaning, humour and metaphors.
- May have a desire to be sociable but have difficulties in understanding non-verbal signals, including facial expressions.
- May have difficulties in forming and maintaining social relationships and interactions.
- May develop 'special interests' that may become obsessional.
- Regulating of personal emotions may be difficult. Often dislikes change and therefore develops daily routines.
- Has good memory and retention of facts but may have difficulties with abstract thoughts.
- Recognition of the written word may be good, but there may be difficulties with comprehension, especially at a high level of literacy where interpretation of subtle meanings is required.
- Written work may be repetitive and not focused.
- Academic work may vary in quality depending on how meaningful or interesting the subject appears to the student.

Possible supportive teaching and learning strategies

- Understand that the student's perspective of the world will differ from that of other students.
- Raise awareness amongst peers of the needs of the student to avoid possible misunderstandings related to social situations and communication.
- Interactive and group skills may need to be explicitly taught.
- Use group strategies that highlight the student's strengths, thus supporting peer acceptance.
- Identify clear rules and expectations in relation to work and behaviour.
- Provide a consistent approach supported by routines, but introduce and teach an approach signal to prepare the student for change.
- Agree on strategies with the student if stress or frustrations overcome him/her.
- While recognising the need for students to discuss their 'special interests', limit the time allowed. When possible, link their interest to the subject being taught; e.g. when exploring landforms, a student obsessed by mountain ranges might be asked to research this particular landform.
- Add additional explanations or clarity when introducing new concepts or ideas, especially if they are abstract, and check for understanding.
- Promote completion and speed of written work through the use of structure (writing frames) and the use of technology.

ASPERGER'S SYNDROME PROFILE

	• Use supplementary visual resources, pictures, maps and photographs relating to the place, concept or theme being discussed. • Frequently direct questions towards the student to engage them in lessons, but allow time for responses. Alert them to this by using their name.
Preparation and planning for outdoor learning	• Provide structure to all aspects of the field trip, i.e. pre-planning, preparation, organisation, task requirements and completion, identification or recording of outcomes and learning. • Give clear instruction on tasks to be completed. • Give clear guidance on suitable behaviour whilst out of school, extended through the use of social stories.

Further information

www.autism.org.uk

Attention deficit disorder/attention deficit hyperactivity disorder (ADD/ADHD)

'Attention deficit disorder' is the term used to describe a neurological based dysfunction which affects performance, self-regulation and impulsivity. Some students may also present hyperactivity. Students demonstrate a normal range of intellectual ability, but the ADHD may in itself cause barriers to learning. All students are different and will present individual patterns of behaviour. Many will display some, if not all, of the characteristics listed here to a degree.

Main characteristics

- short attention span; difficulty in following instructions and completing tasks
- easily distracted by noise, movement of others, objects that attract attention
- often doesn't listen when spoken to
- fidgets and becomes restless, can't sit still
- interferes with other students' work
- can't stop talking, interrupts others, calls out
- runs about when inappropriate
- has difficulty waiting or taking turns
- acts impulsively without thinking about the consequences

ADD/ADHD PROFILE

Potential impact of characteristics on learning	• Talkative, energetic, enthusiastic, inquisitive, impulsive and adventurous. • Easily distracted, cannot maintain attention or refocus on tasks, forgetful, disorganised and has poor attention to detail. • Slow to complete tasks/work set. • Acts and speaks without thinking, interrupts, may have a low frustration and tolerance threshold. • May appear dreamy and distant or restless, fidgety, talkative, and have difficulty remaining in seat, waiting their turn. • Will continue to discuss or argue and will not let issues drop. • Misunderstands social situations, demands their own way. • Poor co-ordination, clumsy, and written work may be messy. • No concept of structuring environment, belongings, school work or projects, loses things, unaware of time. • May demonstrate mood swings. • May experience specific learning difficulties or underachieve.

ADD/ADHD PROFILE

Possible supportive teaching and learning strategies	• Accept the student for who they are and acknowledge their good points.
	• Develop a trusting and positive relationship with the student that promotes the student's self-esteem.
	• Link the geography curriculum to real-life experiences and real-life applications.
	• Structure the environment and reduce clutter; sit the student away from busy (distracting) classroom areas. Make sure that maps and other graphical representations are clear and that only essential information is included.
	• List main learning objectives, ideas, concepts and tasks to be completed at the beginning of lessons. Break tasks down into small steps.
	• Give clear directions orally and visually and where possible provide a model of what to do.
	• Provide breaks within longer tasks and/or alternative lesson activities.
	• Find alternative methods other than tests to identify what students know or have learnt, such as self-assessment and peer review.
	• Provide opportunities for the student to move about legitimately, i.e. take a message or vary activities within the lesson. Allow 'fidget toys' where appropriate (with strict rules on their use).
	• Help students to use written lists that identify personal responsibilities, resource requirements, tasks, activities etc.
	• Have clear and realistic behaviour expectations that are expressed in positive terms.
	• Tell the student well in advance if changes are to take place.
	• Do not enter into in-depth confrontations or discussions; rather, deflect the student to a previous activity that they have found calming.
	• Provide visual resources alongside verbal explanations and lists to support the organisation of tasks.
	• Repeat and paraphrase key concepts in a variety of ways.
	• When asking questions allow the student thinking time before requesting a response.
Preparation and planning for outdoor learning	• Make learning objectives clear.
	• Use co-operative groupings.
	• Make behavioural expectations and rules explicit.

Further information

ADDIS	www.addiss.co.uk
ADHD Foundation	www.adhdfoundation.org.uk
Young Minds	www.youngminds.org.uk

Autistic spectrum disorder (ASD)

Students with ASD have difficulties in social interaction, communication and areas that involve imagination and flexibility of thought. This is known as the 'Triad of Impairments'. Students with ASD cover the full range of ability, and the severity of the impairment varies widely. Some students also have learning disabilities or other difficulties. Four times as many boys as girls are diagnosed with ASD.

Main characteristics

- Social interaction
  Students with ASD find it difficult to understand social behaviour, and this affects their ability to interact with children and adults. They do not always understand social contexts. They may experience high levels of stress and anxiety in settings that do not meet their needs or when routines are changed. This is particularly significant when planning out-of-classroom activities, since this change can lead to inappropriate behaviour. The use of social stories that rehearse situations prior to the event is helpful here. A department may wish to create a bank of such stories that consider appropriate behaviour in fieldwork situations.

- Social communication
  Students with ASD have difficulty understanding the communication of others and developing effective communication themselves. Often their understanding of the use of non-verbal and verbal communication is impaired. They have a literal understanding of language. Many are delayed in learning to speak, and some never develop speech at all.

- Social imagination and flexibility of thought
  Students with ASD have difficulty in thinking and behaving flexibly which may result in restricted, obsessional or repetitive activities. They are often more interested in objects than people, and may have intense interests in one particular area, such as trains or vacuum cleaners. Students work best when they have a routine. Unexpected changes will cause distress. Some students with ASD have a different perception of sounds, sights, smell, touch and taste, and this can affect their response to these sensations.

AUTISTIC SPECTRUM DISORDER PROFILE

Potential impact of characteristics on learning	<ul><li>Can sometimes 'zone out' in lessons and not hear instructions.</li><li>Does not always understand conceptual expressions or phrases, and misunderstands what is being asked of him/her.</li><li>The student needs things to be concrete or visual to make sense so doesn't always understand questions.</li></ul>

AUTISTIC SPECTRUM DISORDER PROFILE

	• Needs teachers to call their name for them to realise they are included in an instruction.
	• Often shows signs of stress when concentrating and is agitated.
Possible supportive teaching and learning strategies	• Give advance warning of any changes to the usual routine, e.g. change of classroom, change of teacher etc.
	• Make expectations very clear: link the task with the learning objectives.
	• Liaise with parents; they may have some useful strategies.
	• Give individual instructions, using the student's name. Avoid using too much eye contact as this can cause distress.
	• Use real examples of places within their experience.
	• Use as many real-life experiences as possible.
	• Provide visual resources in class: objects, artefacts, pictures, picture dictionaries.
	• Place student in a small group or a pair for collaborative work.
	• Use computers. Students with autism can have unusual fixations on parts of objects. They may focus on objects as though through a tunnel. When using a computer this can mean that they are able to focus totally on the screen and to block out all distractions around them. Using a computer can provide a secure comfortable environment as they feel more in control of their surroundings. They are less likely to fail, and they can choose whether or not to communicate with others.
Preparation and planning for outdoor learning	• Give clear instruction on tasks to be completed.
	• Use a digital camera or video to record appropriate behaviour. Put the photos into a book and add a sentence to each one: 'Here I am using a clipboard carefully'. Read through this book before each field trip or out-of-classroom experience.
	• Give clear guidance on suitable behaviour whilst out of school, extended through the use of social stories.

Further information

The National Autistic Society	www.autism.org.uk
Autism Education trust	www.autismeducationtrust.org.uk

Cerebral palsy (CP)

Cerebral palsy is a persistent disorder of movement and posture. It is caused by damage or lack of development to part of the brain before or during birth or in early childhood. Problems vary from slight clumsiness to more severe lack of control of movements. Students with CP may also have learning difficulties. They may use a wheelchair or other mobility aid.

Main characteristics

There are three main forms of cerebral palsy:

- Spastic cerebral palsy – associated with stiff or tight muscle tone resulting in a decreased range of movement. This stiffening of muscle tone can be very painful and affect different parts of the body.
- Dyskinetic cerebral palsy – sustained or intermittent involuntary muscle contractions often affecting the whole body.
- Ataxic cerebral palsy – an inability to activate the correct pattern of muscles during movement, resulting in an unsteady gait with balance difficulties and poor spatial awareness.

Pupils with CP will often have a combination of the different forms and may also have communication difficulties. The following terms are used to denote the extent of a person's cerebral palsy:

- 'Hemiplegia' means the person is affected on one side of the body.
- 'Diplegia' means that two limbs are affected.
- 'Monoplegia' is where one limb is affected.
- 'Quadriplegia' is where all four limbs are affected.

PHYSICAL DISABILITY PROFILE

Possible impact of characteristics on learning	The impact on learning will vary in relation to the degree of the disability, any associated medical conditions and/or the degree of learning difficulty but may include some of the following: • Difficulties with mobility, movement, maintaining posture, fine motor co-ordination and skills. • Poor attendance due to illness or medical and therapeutic appointments. • Poor attention, concentration, organisational and memory skills. • Easily distracted. • Spatial or positional perception problems. • Co-ordination of visual motor skills. • Speech and communication difficulties.

PHYSICAL DISABILITY PROFILE

	• Poor basic skills. • Difficulties in the physical process of writing. • Poor confidence and self-esteem. • Fatigue.
Possible supportive teaching and learning strategies	• If appropriate, seek the advice of other professionals, i.e. physiotherapist, occupational and/or speech therapist. • Structure the environment by ensuring that appropriate resources are available to provide a good physical seating and working position. • Investigate the need for additional resources to support the physical aspect of writing and alternative approaches to recording. • Build on prior knowledge; reinforce abstract concepts with supplementary visual resources and concrete materials. • Reinforce new skills and generalisation by using a range of methods in different settings, and use repetition. • Check understanding through questions. • Support work using material from the board by highlighting key information. The student may find it useful to make a voice recording, or an adult may act as scribe if appropriate. • Allow additional time to meet physical and personal needs and to complete tasks and activities. • Address the student directly, not through the TA.
Preparation and planning for outdoor learning	• Discuss with the student, parents and school nurse any possible issues that may arise. Produce a care plan to address specific physical or medical needs. • Pre-visit the location to ensure access and appropriate facilities. • Have staff pre-read handouts and outline tasks to be accomplished before fieldwork. • Provide adapted recording sheets or alternative forms of recording. • Allow extra time for completion of tasks. • Organise work so that the student forms part of paired or group activity, and differentiate tasks within the group.

Further information

Scope
0808 800 3333
www.scope.org.uk

Down's syndrome (DS)

DS is a genetic condition caused by the presence of an extra chromosome 21. People with DS have varying degrees of learning difficulties ranging from mild to severe. They have a specific learning profile with characteristic strengths and weaknesses. All share certain physical characteristics but will also inherit family traits, in physical features and personality. They may have additional sight, hearing, respiratory and heart problems.

Main characteristics

- delayed motor skills
- takes longer to learn and consolidate new skills
- limited concentration
- difficulties with generalisation, thinking and reasoning
- sequencing difficulties
- stronger visual than aural skills
- better social than academic skills

DOWN'S SYNDROME PROFILE

Potential impact of characteristics on learning	• Poor co-ordination that may impact on physical activities and writing skills. • Smaller vocabulary, articulation problems, difficulties in learning the rules of grammar. • Difficulties with processing, understanding, assimilating and responding to spoken language. • Difficulties in understanding complex verbal instructions. • Poor problem-solving and generalisation of knowledge, understanding and skills. • Difficulties in processing, retaining, organising, consolidating and sequencing information. • Difficulties in recording information, e.g. selecting, organising and sequencing relevant information. Short concentration span.
Possible supportive teaching and learning strategies	• Investigate additional resources to support the physical aspect of writing and alternative approaches to recording. • If appropriate, seek the advice of other professionals, i.e. physiotherapist, occupational and/or speech therapist. • Check understanding through questions. • Build on prior knowledge, and reinforce abstract concepts with supplementary visual resources and artefacts relating to the place, theme or concept. • Reinforce new skills and generalisation by using a range of methods in different settings and by offering opportunities for repetition.

DOWN'S SYNDROME PROFILE

	• Support note taking by highlighting key information, or by using scaffolds and writing frames. • You may also find the use of key word banks, visual resources and sentence sequencing cards helpful to support recording. Consider software such as 'Clicker'. • Provide a supportive structure by teaching routines and rules explicitly, and warn in advance of any changes that may occur. • Ensure student's position in class takes into consideration visual or auditory difficulties. • Reinforce speech through visual resources and/or signs. • If required, set differentiated tasks linked to the work of the class.
Preparation and planning for outdoor learning	• Provide clear sequential instructions of events and task requirements prior to field study. • Provide a structure for recording through writing frames or scaffolds. • Promote paired or group work. • Review the terrain in order to assess accessibility and/or additional support requirements.

Further information

The Down's Syndrome Association
020 8682400
www.downs-syndrome.org.uk

Foetal alcohol syndrome

'Foetal alcohol syndrome' (FAS) and 'foetal alcohol spectrum disorders' (FASD) are umbrella terms for diagnoses relating to a child's exposure to alcohol before birth. Alcohol can affect the development of all cells and organs, but it is the brain and nervous system that are particularly vulnerable. Each person with FAS/D may face a range of difficulties across a spectrum from mild to severe.

Main characteristics

- visual impairment
- sleep problems
- speech and language delay
- impulsivity and/or hyperactivity
- memory problems
- inappropriate social behaviour

FOETAL ALCOHOL SYNDROME PROFILE

Possible supportive teaching and learning strategies	• Gather information from parents and other professionals involved with the pupil to find the most effective ways of teaching him/her (perhaps through the SENCO in the first instance). • Find out the pupil's strengths, and use these as starting points for learning. • Keep instructions simple, and offer information in verbal and visual form, supported by mime, gesture and facial expression. • Ensure class routines are explicit and followed consistently. • Use concrete and positive language, e.g. 'Walk' rather than 'Don't run'. • Check that the pupil knows and understands any school or class rules. • Specify clearly what is expected for any task or activity. • Provide a memory mat or audio recording facilities to support retention of information, e.g. homework tasks, spelling, etc.

Further information

www.drinkaware.co.uk/alcohol-facts/health-effects-of-alcohol/fertility-and-pregnancy/foetal-alcohol-syndrome-fas/

Fragile X

Fragile X is a complex genetic condition and is one of the most common known causes of inherited learning difficulties. It occurs because of an abnormality on the long arm of the X chromosome that can cause a variety of intellectual, behavioural and physical differences. Fragile X affects both males and females, although it is often the case that males exhibit more severe symptoms than females. Cognitive and learning abilities may range across the whole spectrum.

Main characteristics

- delayed and disordered speech and language development
- difficulties with the social use of language
- articulation and/or fluency difficulties
- verbal skills better developed than reasoning skills
- repetitive or obsessive behaviour such as hand-flapping, chewing etc.
- clumsiness and fine motor co-ordination problems
- attention deficit and hyperactivity
- easily anxious or overwhelmed in busy environments

FRAGILE X PROFILE

Potential impact of characteristics on learning	• Has difficulties in attending, remaining on task and transferring from one task to another. Has difficulties with problem-solving, thinking and reasoning skills and abstract learning tasks.
	• Sees tasks as a whole rather than their component parts.
	• Is likely to be a visual learner.
	• Has social anxiety, which often inhibits a student's ability to communicate effectively.
	• May find cause and effect and open-ended questions difficult.
	• May present difficulties in social skills and adaptation of such skills in new settings.
	• May present repetitive and sometimes unusual behaviour, e.g. hand-flapping, and demonstrate over-activity, social anxiety, inflexibility and an insistence on routine.
	• Is susceptible to stimulus overload and busy environments.
Possible supportive teaching and learning strategies	• Teach group and social skills explicitly.
	• Provide timetables and lists to support sequences, transfer of learning and change.
	• For many students with SEN, we tend to break tasks down into smaller steps. However, for some students with Fragile X it may be more appropriate to use the opposite approach, i.e. to demonstrate the whole and then the individual components.

FRAGILE X PROFILE

	• Provide concrete examples, and/or model task or activity whenever possible.
	• Use varied activities and tasks, and keep them short.
	• Give clear indications throughout the structure of the task about what is expected, what has to be done and when it is completed.
	• Use alternative ways of recording, e.g. computer or adapted worksheets that are free from clutter.
	• Provide a seating arrangement which avoids busy areas of the classroom and allows the student not to feel crowded or that their personal space is being invaded.
	• Agree with the student strategies to manage over-stimulation, frustration etc.
	• Provide visual information (including maps and photographs) that is simplified to eliminate clutter or excessive stimulation.
	• Provide objects, artefacts and visual resources to support learning.
	• Whenever possible, relate learning to the student's everyday experiences.
	• Provide visual resources to help the sequencing and organising of tasks, e.g. arrows, numbers etc.
Preparation and planning for outdoor learning	• Provide structured materials, worksheets and visual resources to clearly identify the purpose of each task, the tasks themselves and the sequence of activities and expected outcomes.
	• Make learning objectives clear.
	• Use supportive or co-operative groupings.
	• Make expectations and rules for behaviour clear.

Further information

Fragile X Society
01424 813147 (Helpline)
01371 875100 (Office)
www.fragilex.org.uk

Learning difficulties/disabilities

The terms 'learning disability' and 'learning difficulty' are used to describe a wide continuum of difficulties ranging from moderate (MLD) to profound and multiple (PMLD). Pupils with learning disabilities find it harder to understand, learn and remember new things, and may have problems across a range of areas such as communicating, being aware of risks and managing everyday tasks.

Moderate learning difficulties (MLD)

The term 'moderate learning difficulties' is used to describe pupils who find it extremely difficult to achieve expected levels of attainment across the curriculum, even with a well-differentiated and flexible approach to teaching. These pupils do not find learning easy and can suffer from low self-esteem and sometime exhibit unacceptable behaviour as a way of avoiding failure. For all pupils with learning disabilities, the social aspect of school is a crucial element in their development and understanding of the 'culture' of young people, so it is important for them to have friends who don't have learning disabilities as well as those who do. As the SEND Code of Practice says at 6.2: "Schools must . . . ensure that children and young people with SEN engage in the activities of the school alongside pupils who do not have SEN" (p. 92).

Main characteristics

- difficulties with reading, writing and comprehension
- inability to understand and retain basic mathematical skills and concepts
- immature social and emotional skills
- limited vocabulary and communication skills
- short attention span
- under-developed co-ordination skills
- lack of logical reasoning
- inability to transfer and apply skills to different situations
- difficulty remembering what has been taught
- difficulty with organising themselves, following a timetable, remembering books and equipment

In geography these students can, if appropriate, be supported by modifying the geography curriculum by:

- choosing material from earlier key stages
- maintaining, reinforcing, consolidating and generalising previous learning, as well as introducing new knowledge, skills and understanding
- focusing on one aspect, or a limited number of aspects, of a topic
- including experiences that let students at early stages of learning gain knowledge and understanding of geography in the context of everyday activities

- helping students experience geography for themselves, using a sensory approach to experience and investigate familiar places, and by contact with different people

There is no requirement to teach geography at Key Stage 4. However, for many students, geography offers satisfying challenges both as a subject in its own right and as a means of developing skills in many other areas of the curriculum.

MODERATE LEARNING DIFFICULTIES PROFILE

Potential impact of characteristics on learning	• Poor reading, writing and numeracy skills. • Problems with processing, retaining and recalling information. • A limited vocabulary, immature use of language and grammatical structures. • Difficulties in understanding and responding to instructions and open-ended questions. • Poor listening and a short attention span. • Limited on-task behaviour. • Poor organisational skills. • Immature social and emotional skills. • Disruptive behaviour through constantly distracting and talking to other students. • The use of a range of avoidance techniques.
Possible supportive teaching and learning strategies	• Provide structure and routine to lesson format. • Break learning down into small steps. • Plan short tasks and varied activities, which include repetition of key concepts and ideas presented. Differentiate in terms of content, resources and outcomes, and use a wide range of methods of communication including speech, images, pictures, charts, diagrams and symbols. • Support learning through the use of concrete examples, e.g. word lists, pictures, photos, symbols, sequencing prompts, shortening of text etc. • Repeat information and instructions in different ways. Use simple language and a range of questioning techniques. • Rehearse with the student what is to be recorded before they begin, and use a range of structures to support reporting and recording, i.e. scaffolding, writing frames. • Provide an environment and physical structure to reduce inappropriate and disruptive behaviours, e.g. sitting away from obvious distractions. • Identify clear rules and expectations. • Identify approaches to the management of disruptive behaviours and avoidance of attention-seeking behaviours, and apply consistently. • Keep listening activities brief. • Catch them being 'good' and reward them immediately.
Preparation and planning for outdoor learning	• Provide a clear outline of activities and learning outcomes for the student prior to field study. • Use visual resources to support printed information that may also be used on the visit itself. • Clearly identify rules and behaviour expectations. • Provide a structured recording format. • Plan group or paired activities.

Severe learning difficulties

Young people with severe learning difficulties may have a complex pattern of educational, social and emotional needs. Many have communication difficulties and/or sensory impairments in addition to more general cognitive difficulties. They can also have difficulties with mobility, co-ordination and perception. Some students may use signs and symbols to support their communication and understanding. (Symbolic and spatial representation of course has great resonance with some of the key characteristics of geography.) They will be achieving well below students of the same age and may remain within the P scale range for most or all of their school life.

Access to mainstream settings for these students varies across local education authorities. In some cases, students attend a mainstream setting on a part-time basis, accompanied by a TA. This can be extremely successful in providing positive role models for the youngster with severe learning difficulties, improving self-esteem and motivation and providing access to specialist subject teachers. These arrangements work well when there is informed dialogue and supportive liaison between special school and mainstream teaching teams.

Main characteristics

Students may experience difficulties with:

- personal independence and self help skills
- 'learning to learn' and memory skills
- expressive communication
- development of basic skills
- problem-solving

SEVERE LEARNING DIFFICULTIES PROFILE

Potential impact of characteristics on learning	• Explicit learning programmes may be required in all areas of development. • Limited incidental learning, i.e. has difficulty picking things up as you go along and making connections. 'Learning to learn' skills such as attention, concentration, on-task and group working skills may need to be specifically taught. • Limited vocabulary. • Limited use of language and comprehension. • Difficulties with problem-solving and responding appropriately to open-ended questions.

	• Difficulties in processing, retaining and recalling information and generalising knowledge, understanding and skills.
	• Poor reading, writing and mathematical skills.
Possible supportive teaching and learning strategies	• Create individual educational programmes to address priority areas of learning.
	• Use achievable tasks for each lesson or module of work.
	• Break learning down into small steps. Allow time for students to process information and formulate responses.
	• Include a series of short, varied activities in each lesson, with clear, specific instructions.
	• Use simple language that is based on the student's understanding but continue to support and extend the student's communication, comprehension, vocabulary and use of language for a range of purposes.
	• Build on prior learning, and introduce new concepts through concrete experiences whenever possible. Set differentiated, practical tasks that link to the work of the rest of the class.
	• Use objects, pictures and symbols relating to the place or theme being explored to support and supplement learning.
	• Devise 'doing activities' that focus on specific areas of the geography curriculum, e.g. sensing different types of ground, e.g. soil, sand, grass and tarmac, to appreciate places are both similar and different.
	• Identify key words and questions for geography, and learn some relevant signs to communicate these appropriately to the student.
	• For each topic, create a set of stimuli that appeal to different senses, e.g. things that can be touched, smelled or tasted.
	• Use annotated photographs to record learning experiences and achievements.
Preparation and planning for outdoor learning	• Provide a preview of location, activities, tasks to be completed and possible learning outcomes, e.g. a video from a previous visit made by other students or a TA.
	• Work with the student and parents to clearly identify organisational and resource requirements.
	• On the day provide resources to support the practical activities using student's preferred communication system.
	• Provide clear behaviour rules, boundaries and realistic expectations.
	• Identify an additional adult to support the student and liaise effectively.

Profound and multiple learning difficulties (PMLD)

Students with profound and multiple learning difficulties present a range of needs in addition to severe learning difficulties. They do not present a homogeneous group. In addition to very severe learning difficulties, students have other significant challenges to face, such as physical disabilities, sensory impairments or severe medical conditions. Students with PMLD require a high level of adult support, both for their learning needs and for their personal care, and are usually placed in specialist settings.

They are able to access the curriculum through sensory experiences and stimulation. Some students communicate by gesture, eye-pointing or symbols, others by very simple language. They may work within P scales P1–P4 for most of their school career. (The P scales provide small, achievable steps to monitor progress.)

Main characteristics

Students may experience difficulties with:

* movement (physical disabilities may require specialised physical resources such as a wheelchair, standing frame or side liner)
* sensory impairment
* complex medical conditions
* inappropriate behaviour
* communication
* memory
* learning beyond that associated with the early stages of development in all areas

PROFOUND AND MULTIPLE LEARNING DIFFICULTIES PROFILE

Potential impact of these characteristics on learning	• Requires a high level of adult support to access learning and meet personal, physical and medical needs. • Has limited movement that may seriously restrict the student's physical access and reaction to their environment. • Relies on observant and vigilant staff to respond to personal methods of communication, i.e. movement of left hand, vocalisation.
Possible supportive teaching and learning strategies	• Individual educational programmes designed by a multi-professional team that provide an holistic approach to teaching and learning. • Break learning down into small, achievable steps. This may well focus on the development of the student's sense of personal geography.

PROFOUND AND MULTIPLE LEARNING DIFFICULTIES PROFILE

	• Ensure a multi-sensory dimension to teaching and learning, e.g. providing sight, sound and taste experiences relating to places, themes or concepts that the class is addressing. • Focus on real experiences in a variety of settings to support a sense of place. • Explicitly teach 'learning to learn' and social skills. • Revisit learning, and extend in different contexts and settings. • Revisit locations and experiences in school through multi-sensory activities. • Consistent approach by all staff. • Supportive and flexible learning environment and timetable to provide access to inclusive as well as specialist learning opportunities or settings.
Preparation and planning for outdoor learning	• Be creative in reviewing what field studies may provide the required learning experiences and opportunities for the students; these may not involve great distances or difficult terrain. • Pre-visit by the teacher to assess physical access, facilities, staff requirements, medical support and any additional risks that the visit may pose. • Communicate with the student throughout the visit about what is happening immediately and what will happen next. • Use aspects of the environment to extend students' knowledge of the location, e.g. sounds, smells etc. of a busy market in contrast to the smell and sounds of a quiet wooded area; use of the school grounds to explore different weather conditions, a cobbled path, a smooth path, rough terrain.

Further information

MENCAP
020 7454 0454
www.mencap.org.uk

Sensory and/or physical impairment

Visual impairment (VI)

There is a wide range of visual impairment in both type and degree. These include conditions of short- and long-sightedness that can be successfully corrected by having pupils wear glasses, monocular vision (vision in one eye) and partial sightedness; at the severe end of the spectrum are those pupils who have virtually no useful sight and are registered as blind. The majority of students who are registered as blind have some sight, but the extent will vary considerably. Students with visual impairment cover the whole ability range.

Main characteristics

These will vary from student to student depending on the effect of their specific visual impairment but may involve the following:

- difficulties with mobility
- poor spatial awareness
- difficulties with social and communication skills
- becoming tired after periods of intense concentration

VISUAL IMPAIRMENT PROFILE	
Potential impact of characteristics on learning	• May require differentiation in terms of curriculum content, access, resources and outcome. • May require additional curriculum, i.e. specialist teaching of Braille, mobility, tactile strategies, keyboard, social and life skills. May be unable to access the curriculum via the visual media used. • Takes longer to process and assimilate information, and requires additional time to complete tasks and activities. • Need additional time for their eyes to adapt to impulses from the environment. • May require additional formats for reading and writing, and access to specialised resources, e.g. Braille. • May become visually fatigued when concentrating for sustained periods of time. • Limited perception of environment. • Poor spatial awareness.
Possible supportive teaching and learning strategies	The type of specialised teaching strategies and resources required will differ from student to student. Expert advice and training must be sought from a specialised teacher. Ask the student to identify his/her requirements. Additional strategies may include some or all of the following: • Ensure optimum seating position for student in terms of light source etc. • Avoid standing with the light behind you when speaking to the class.

VISUAL IMPAIRMENT PROFILE

| | • Ensure safe access to all areas, materials and resources; clearly identify any changes that have taken place, and walk the student through the new layout if necessary; ensure that the floor is free of clutter and that anything hanging from the ceiling is not an obstacle.
• Use a multi-sensory approach whenever possible. Use images, sounds, tastes and smells.
• If appropriate, adapt material and resources in terms of size, clarity and contrast of print against background.
• Allow time for the student to relax following periods of intense concentration.
• Provide only the information required, avoiding clutter and unnecessary details.
• Provide oral or written descriptions, or simple drawings to support photographs etc.
• Use objects, artefacts, tactile displays, pictures, graphs and models.
• Produce tape recorded descriptions of regularly used maps and photos.
• Liaise with the local university geography department for use of raised, simplified line maps.
• Use models for landforms. |
| Preparation and planning for outdoor learning | • Discuss with students, parents and specialist teachers for the visually impaired to identify possible issues and solutions including any additional staff support needed if appropriate.
• Pre-visit essential. |

Further information

Royal National Institute for Blind People (RNIB)
0303 123 9999
www.rnib.org.uk

Hearing impairment (HI)

Hearing impairment is a broad term that is applied to partial or complete loss of the ability to hear. The loss may vary in both nature and severity, described as mild, moderate, severe or profound. An impairment may occur through a conductive loss, which is usually temporary and ranges from mild to moderate, or be 'sensorineural', which is permanent hearing loss. Students may use lip reading and hearing aids, and those with partial hearing loss may still be able to communicate effectively orally. Students whose degree of loss is complete may use a variety of alternative communication language forms such as British Sign Language. Students with a hearing impairment cover the whole ability range.

Main characteristics

These will vary from student to student depending on the effect of their specific hearing impairment but may involve the following:

- use of a variety of communication methods
- good visual skills
- difficulties in interactional and social skills
- emotional difficulties and poor self-esteem that can lead to feeling isolated and withdrawn
- becoming tired and fatigued after periods of intense concentration

---

HEARING IMPAIRMENT PROFILE

Potential impact of characteristics on learning	• May have difficulties with language structure and literacy. • May miss important information. • May find it difficult to participate in or follow group discussions. • Requires more time to assimilate and respond to information, questions etc. • Will become fatigued when lip reading as this requires intense and sustained concentration. • May require an interpreter or note taker. • Environment needs to be light and as free from background noise as possible.
Possible supportive teaching and learning strategies	The type of specialised teaching strategies and resources required will differ from student to student. Expert advice and training must be sought from a specialised teacher. Ask the student to identify his/her requirements. Additional strategies may include some or all of the following: • Obtain the student's attention before you speak. • Speak clearly and at a normal pace; do not shout. • Support the student's ability to lip read by not covering your mouth with your hand, pen etc. or putting your back towards the student. Avoid standing so the light is behind you and your face is in shadow. • Do not talk while writing on the board. • Rephrase or repeat information if required. • Provide additional information in written or visual form. • If the students are using an interpreter or note taker, always speak directly to the student.

	• In group discussions sit students in a circle, and identify a set of rules and cues to manage the interactions; e.g. each person in turn makes a contribution.
	• Clearly outline the lesson focus, activities and expected learning outcomes and support in writing.
	• If possible, provide information prior to lessons. Supplement oral or signed communications with written or visual material.
	• Use short sentences; they are easier and quicker to comprehend than long ones.
	• When possible, begin explanations with concrete examples before moving on to more abstract ones.
	• Present only one source of visual information at a time.
	• Maximise the use of visual media.
	• At every opportunity obtain feedback from the student to determine understanding.
Preparation and planning for out-door learning	• Provide step-by-step information regarding organisation, requirements, focus of study, activities and expected outcomes prior to the visit.
	• Provide peer or group support to take notes – it is impossible to write and lip read at the same time.

Further information

Action on Hearing Loss 020 7296 8000 www.actiononhearingloss.org.uk
The National Deaf Children's Society 020 7490 8656 www.ndcs.org.uk

Multi-sensory impairment (MSI)

The term 'multi-sensory' is used to represent a diverse group of students who may have varying degrees of visual and hearing impairment, which may be combined with other disabilities. The complexity of their needs may make it difficult to ascertain a true picture of their intellectual abilities. Dual sensory impairment may have a significant impact on the student's ability to gain information, meaning and understanding from and about their environment and may result in sensory deprivation. Teaching approaches must facilitate the effective use of all the senses, including residual hearing and vision, within an environment that ensures the development of trusting relationships and a consistency of approach.

Main characteristics

These will vary considerably from student to student but may include some of the following:

- visual and hearing impairment
- additional disabilities
- idiosyncratic methods of communication
- challenging or self-injurious behaviour
- isolated and withdrawn behaviour

MULTI-SENSORY IMPAIRMENT PROFILE

Possible impact of characteristics on learning	• Restricted information through vision and hearing. • Access to incomplete or distorted perceptions of the world as a result of sensory limitations. • Inability to sequence and link activities. • Aversion to contact with aspects of their environment including people. • Difficulties in establishing trust and developing relationships. • Difficulties in accessing, making sense of and processing information. • Does not pick up information incidentally. • Becomes tired and fatigued after periods of intense concentration.
Potential supportive teaching and learning strategies	• Calm atmosphere. • Clear physical structure which provides safe pathways for mobility. • Identify different areas and details of the environment, and link these with regular activities. • Use matt surfaces to avoid glare. • Ensure effective and appropriate lighting levels. • Ensure that the environment and working areas are free from clutter. • Use furnishings to control noise levels and distractions. • Use blinds and curtains to control glare from windows. • Each member of staff has a badge coloured or with a symbol, to identify who they are to the student. The student is alerted to the adult's presence and to when the adult leaves. • Provide consistent approaches through routines and responses of staff. • Give students opportunities to control and act upon their environment and make choices. • Allow the student time to respond. • Respond consistently to the student's communicative behaviour. • Provide frequent opportunities for communication to take place.

	• Ensure there is commentary to inform the student what is going to happen, what will happen next and when an activity is finished. • Break learning down into small steps, but teach them as part of the whole activity. Concrete examples should be used. • Revisit the experience through simulated sensory activities.
Preparation and planning for outdoor learning	• Pre-visit the location to ensure access and facilities are appropriate. • If appropriate, rehearse the visit in familiar surroundings or through simulated sensory experience.

Social, emotional and mental health needs

This area includes pupils who experience a wide range of difficulties characterised in a number of ways, including becoming withdrawn or exhibiting behavioural difficulties. Behaviours such as these may reflect underlying mental health difficulties including depression, anxiety and eating disorders. These difficulties can be seen across the whole ability range and have a continuum of severity. Attachment disorders and ADD will also be part of this continuum. Pupils with special educational needs in this area are those who have persistent difficulties despite the school having in place an effective school behaviour policy and a robust personal and social curriculum.

Main characteristics

- inattentive, poor concentration and lack of interest in school/school work
- easily frustrated, anxious about changes
- difficulty with working in groups
- unable to work independently, constantly seeking help
- confrontational – verbally aggressive towards students and/or adults
- physically aggressive towards students and/or adults
- destroys property – their own/others
- appears withdrawn, distressed, unhappy, sulky; may self-harm
- lacks confidence, may act extremely frightened, lacks self-esteem
- finds it difficult to communicate and to accept praise

SOCIAL, EMOTIONAL AND MENTAL HEALTH NEEDS PROFILE
SOCIAL DIFFICULTY PROFILE

Potential impact of characteristics on learning	• Demonstrates a lack of interest in school or learning. • Underachieves. • Has poor attendance. • Finds it difficult to concentrate and complete tasks; works slowly. • Often displays inappropriate behaviour that disrupts the lesson, is harmful to self and others or is extremely isolated, withdrawn, distressed and unresponsive. • Demonstrates mood swings and is easily frustrated or emotional. • Finds it difficult to work co-operatively with others and to share equipment or space. • May find it difficult to accept rules or what is perceived as authoritarian figures and may resort to confrontation or argument. • Fails to follow instructions. • Exhibits poor organisation. • May destroy own work or that of others. • Finds it difficult to communicate, trust, interact and form relationships with adults and peers.

SOCIAL, EMOTIONAL AND MENTAL HEALTH NEEDS PROFILE
SOCIAL DIFFICULTY PROFILE

Possible supportive teaching and learning strategies	• Differentiate work, content and outcome to meet the student's learning needs. • List the main learning objectives, ideas, concepts and tasks to be completed at the beginning of lessons. Make clear some of the big ideas in geography to the students. Provide a calm, structured environment with clear routines and procedures. • With the school's behaviour policies and procedures clearly identify behaviour expectations, rules, routines and responsibilities. • Have a small number of important rules and apply them consistently rather than having many that are unmanageable. • Identify clearly the consequences of inappropriate behaviour, and implement fairly and consistently. • Implement, monitor and review individual plans continually. • Recognise and acknowledge positive behaviour in a way that is appropriate to the student, and identify ways to enhance their self-esteem. • Tell students what you want them to do, e.g. 'I need you to . . .', rather than asking, e.g. 'Will you . . .', which provides the opportunity for the negative response 'No!' • Use strong, positive 'I' statements, e.g. 'I care', 'I'd prefer', 'I'd be happier if . . .' • Choose appropriate times for discussion and negotiation, e.g. not in full view of others. If confrontation occurs, stay calm, lower your voice. • Do not enter into in-depth confrontations or discussions; deflect the student, suggest a next step or give choices. • Focus on the behaviour and not the student.
Preparation and planning for field studies	• Make behavioural expectations and rules explicit. • Identify clearly the dangers in specific locations. • Ensure adequate staffing.

Further information

SEBDA is the name for the Association of Workers for Children with Emotional and Behavioural Difficulties
01233 622958
www.sebda.org

Specific learning difficulties (SpLD)

The term 'specific learning difficulties' covers dyslexia, dyscalculia and dyspraxia.

Dyslexia

The term 'dyslexia' is used to describe difficulties that affect the ability to learn to read, write and/or spell stemming from a difficulty in processing the sounds in words. Although found across a whole range of ability, pupils with dyslexia often have strengths in reasoning and in visual and creative skills, but their particular difficulties can result in underachievement in school. While pupils can learn strategies to manage the effects of dyslexia, it is a life-long condition, and its effects may be amplified at times of stress or in unfamiliar situations.

Main characteristics

- The pupil may frequently lose their place while reading, make a lot of errors with high frequency words, have difficulty reading names and have difficulty blending sounds and segmenting words. Reading requires a great deal of effort and concentration.
- Written work may seem messy with crossing outs; similarly shaped letters may be confused, such as b/d/p/q, m/w and n/u; and letters in words may be jumbled, such as 'tired'/'tried'. Spelling difficulties often persist into adult life, and these students may become reluctant writers without appropriate support.
- Personal organisation can be poorly developed.

DYSLEXIA PROFILE	
Potential impact of characteristics on learning	• Loses place when moving from one line to the next; print often becomes blurred, or letters and words appear to move.
	• Finds it difficult to obtain meaning from text and misreads questions.
	• Finds it very difficult to work from the board.
	• Finds it difficult to listen for long periods of time.
	• Becomes stressed and frustrated when unable to write legibly, quickly or neatly and the content does not relate to knowledge.
	• Has difficulty in remembering daily/weekly timetables, instructions and facts for tests and examinations.
	• Rarely completes work on time.
	• Suffers from eye discomfort and headaches and is often tired.
	• Is creative and imaginative.
Possible supportive teaching and learning strategies	• When possible, use a multi-sensory approach; draw on supplementary visual, tactile and auditory resources in addition to text.

- It often helps if the first words of each sentence are highlighted in colour and work is left on the board for sufficient time. Electronic copies of handouts mean that worksheets can easily be altered; the use of PowerPoint on an interactive whiteboard can also be helpful here.
- If place-based websites are being used, it may be worth exploring if the text colours on these can be altered.
- Students with dyslexia may prefer to communicate verbally what they have learnt.
- Use handouts in addition to the whiteboard (on which it is best to use coloured pens; try blue rather than black) that are differentiated in terms of readability: levels of text, number of sentences on each page, font type and size, increased spacing between words and lines, colour of paper (pastels are often better than white). The use of recycled paper can create a geographical discussion as well as reduce glare.
- Remember also to review the text used on maps and diagrams and other forms of visual presentation used in geography. When a handout is text-rich, consider limiting the amount of information and presenting it in a vertical rather than linear format.
- TAs can help the pupil to pre-read text and prepare for the lesson.
- Use of mind-mapping techniques may be helpful.
- Teaching of specific strategies may be helpful, including alternative methods of recording; spreadsheets can provide a visual aid as students with dyslexia can be taught visual methods of laying out their work. Keep a geography dictionary to support an understanding of new words and their meaning.
- Only ask the student to read aloud if you are certain that it is a manageable task; allow for some preparation if possible.
- Create high expectations and a low stress atmosphere. Provide a framework for the lesson. Be specific and try to link details and examples back to the overall concept; for example, when considering coasts and erosion, link place studies such as Studland in Dorset to some of the big ideas, for example, erosion and the notion of a dynamic earth.

Preparation and planning for outdoor learning

In additional to the points already mentioned, consider:
- The amount of time allowed to complete written observations. (Or can they be recorded?)
- The use of drafting, editing and publishing technology. Digital images and video relating to new fieldwork locations can provide helpful support, facilitating access outside of normal lesson time to new and unfamiliar places.
- A dyslexic student, when asked about their fieldwork experiences, stated, 'Fieldwork is the best thing about geography, a good way to learn, but it's difficult for me to make notes in the field. Please put more information in the handouts and remember that it is hard for me to read and write in bright sunlight.'

Further information

The British Dyslexia Association	www.bdadyslexia.org.uk
Dyslexia Action	www.dyslexiaaction.org.uk

Dyscalculia

There are a number of views relating to whether dyscalculia and dyslexia are interrelated or discrete specific learning difficulties. As it is beyond the scope of this chapter to debate these issues, the information given here identifies some basic characteristics of dyscalculia and ways they may impact on teaching and learning in geography.

The term 'dyscalculia' refers specifically to difficulties in acquiring mathematical skills and developing number concepts, facts and procedures.

Main characteristics

- The pupil may have difficulty counting by rote, writing or reading numbers; they may miss or reverse numbers, have difficulty with mental maths and be unable to remember concepts, rules and formulae.
- In maths-based concepts, the student may have difficulty with money, telling the time, directions, and right and left, which are all key abilities to analysing data in the field or outside of the classroom.

DYSCALCULIA PROFILE

Potential impact of characteristics on learning	When using mathematical concepts in relation to geography, students may find it difficult to: • Remember rules, formulae, order of calculations and basic addition, subtraction, multiplication and division facts. • Read, write and record in mathematics figures and facts without making mistakes. • Understand the technical language of mathematics even though they may understand the words in other contexts. • Remember the 'layout' of things and geographical locations. • Demonstrate a sense of direction; they may easily become disorientated in new situations.
Possible supportive teaching and learning strategies	• Whenever possible, use concrete rather than abstract mathematical examples to illustrate a problem; for example, use pictures/models when dealing with traffic data.

DYSCALCULIA PROFILE

- Prepare a prompt sheet electronically, with worked examples of regularly used mathematical tools that are used and applied in geography. Students can refer to this at any time.
- Work through problems with the student, first breaking them down into small steps and allowing time for checking. This is particularly significant for students when they come to undertake their GCSE coursework. Ask for a TA to work alongside students during this time, and carefully consider how to enable peer support whilst both gathering data in the field and analysing it back in school. Seek advice from the staff and TAs who work with the maths department, so that consistent support can be given.
- Provide supplementary visual resources to support the mathematical processes or rules. Use pictures, photographs and models whenever possible. When using maps and scale for decision-making, cardboard templates (drawn to scale) can be used with the map; e.g. when siting a new development, the proposed shopping centre is produced on a cardboard template to scale to support the student in locational decision-making.
- Support the use of a calculator. Give the student extra time to undertake coursework tasks (explain to peers so that they understand that this is fair and reasonable).
- Always have rough paper available for working out.
- Give extra practice on data related tasks, including supplementary activities that involve counting objects rather than just dealing with numbers.

Preparation and planning for outdoor learning

- Review planning and tasks with students prior to the visit; identify possible issues, and provide adapted material to address these.
- Provide reinforcement for work on left/right, map work and directions prior to any fieldwork experience.
- Do not leave students on their own in the field – a peer buddy system operates. This is to support them with their sense of direction. These students have their own strengths too, so careful thought needs to be given to the groups that they work with in order that they help each other.

Further information

The British Dyslexia Association   www.bdadyslexia.org.uk/dyslexic/dyscalculia

Dyspraxia (SpLD)

Generally, 'dyspraxia' is defined as an immaturity of the organisation of movement which includes both fine and gross motor skills. Some students may also have difficulties with language and perception.

Main characteristics

- difficulty in co-ordinating movements; may appear awkward and clumsy
- difficulty with handwriting and drawing, throwing and catching
- difficulty following sequential events, e.g. multiple instructions
- may misinterpret situations, take things literally
- limited social skills, resulting in frustration and irritation
- some articulation difficulties (verbal dyspraxia)

DYSPRAXIA PROFILE

Potential impact of characteristics on learning	• Does not work well in an unstructured environment. • Finds organising equipment and managing time and finishing tasks hard. • Finds it difficult to sit still and is easily distracted. • Finds it difficult to transfer information from the board or to complete a complex sequence of instructions. • Becomes confused with locations and directions, e.g. left and right, up and down, east and west etc. • Can become easily distressed and frustrated.
Possible supportive teaching and learning strategies	• Provide structure: writing frames may be helpful. • Seating position in class should enable the student to view the teacher directly, close enough to hear and see instructions easily; avoid highly stimulating areas, e.g. next to a window or door. • Consult a physiotherapist and/or occupational therapist for appropriate sitting/writing positions or additional resource requirements. It often helps to have both feet resting on the floor, the desk at elbow height and, ideally, a sloping surface to work on. • Break tasks and instructions down into small steps. • Provide handouts, graphs etc. to support lesson content. • Help the student to develop strategies to support organisational skills, e.g. checklists for resources required for specific tasks. • Organise buddies for oral work. • Check understanding verbally. • Use alternative methods to reduce handwriting requirements. • Supplement maps with pictures/photographs and oral interpretation.

DYSPRAXIA PROFILE	
Preparation and planning for outdoor learning	• Provide clear sequential instructions for events and task requirements prior to field study. • Provide structure for recording through writing frames. • Review the terrain in order to assess accessibility and/ or additional support requirements.

Further information

Dyspraxia Foundation
01462 455016
www.dyspraxiafoundation.org.uk

Speech, language and communication needs (SLCN)

The development of language is a complex process that relies on not only the development of organs or muscles that enable or support the production of speech but also the ability to use (expressive) and understand (receptive) language. Understanding of the nuances in verbal and non-verbal communication and social interaction enables us to add to or change the meaning of the words that we use. Difficulties in the development of speech, language and communication may involve any aspect of this process, may cover the full continuum of severity and may last for a short or long period of time. If the difficulties are not related to any other SEND, they are identified as specific language difficulties. It is not difficult to anticipate the impact of specific language difficulties on the learning, social and emotional development of any student.

Main characteristics

General

- specific disability in some aspect of speech, language or communication
- poor interactional and social skills
- emotional difficulties and poor self-esteem
- poor attention and listening skills
- difficulty in assimilating and generalising new concepts and skills and the associated language

Expressive language

- phonology – delay in, or disorder of, speech sound systems
- semantics – difficulty in conveying meaning to others
- pragmatics – difficulties in understanding the rules of communicative interactions
- syntax and grammar – limited use of complex sentence structures

Receptive/comprehensive difficulties

- semantics – difficulty in understanding the meaning of language
- grammar and syntax – difficulties in understanding specific sentence structures
- sequencing – difficulties in the sequencing and organisation of words in sentences and sequencing of information
- poor language memory and retention skills

Potential impact of these characteristics on learning	• Poor literacy skills. • Difficulty in developing and maintaining friendships and understanding social rules. • May become anxious or frustrated; may underachieve. • Poor self-confidence and self-esteem. • Intense concentration leads to student being tired and exhausted by the end of the day. • Difficulty in organising self, environment and materials and following timetables, class routines or rules. • Poor articulation and intelligibility of speech. • Limited use of more complex sentence structures. • Difficulties in the use of phonics, impacting on reading and literacy skills. • Difficulty in communicating or conveying meaning to others. • May use incorrect vocabulary. • Information lacks detail. • Loses focus or subject of conversation. • In communicative exchanges may not maintain eye contact, take turns or use language appropriately for purpose or settings. • Speech errors, e.g. irregular plurals, tenses. • Unable to sequence, plan or organise information or ideas in a co-ordinated and logical way. • Difficulties in remembering and retrieving vocabulary, or may use incorrect label, e.g. using 'cup' when they mean 'spoon'.
Possible supportive teaching and learning strategies	• Work with students, parents and speech therapists to develop an individual education programme based on the student's individual need or to identify specific teaching strategies and resources. • Use visual resources, pictures, symbols and computer programmes to support communication and language skills. • Support lessons and introduction of new subject-specific concepts and vocabulary through concrete experiences and the use of visual resources. • Explicitly teach social, interactional, communication and co-operative working skills as a normal part of lesson structure and group work. • Use lists, subject vocabulary banks, sequencing cards, adapted worksheets, augmentative writing systems etc. • Model task and activities and use of equipment. • Use recorded descriptions of frequently used maps and photographs. • Use a range of strategies to problem-solve, explore ideas and plan, review and record what they know and have learnt, e.g. mind maps, computer overlays etc.

SPEECH, LANGUAGE AND COMMUNICATION NEEDS PROFILE

	• Alert the student that you are communicating with them, directing information by calling out their name. • Clearly outline the lesson objective, activities and learning outcome; you may find that repetition at each stage of the process is helpful.
Preparation and planning for outdoor learning	• Clearly outline objectives, activities and learning outcomes prior to the visit, and discuss with the student possible additional resources and recording requirements. • Support organisational skills through schedules and a timetable. • Plan for group activities, and review communication and language skills to be used. • Make explicit the rules and expectations.

Further information

ICAN	0845 225 4073	www.ican.org.uk
AFASIC	0300 666 9410	www.afasic.org.uk

Tourette's syndrome

Tourette's syndrome is a neurological disorder which is characterised by 'tics'. Tics may be simple or complex and consist of involuntary, rapid or sudden movements and uncontrollable vocalisation which is repeated over and over again. Students may sometimes suppress their tics for a short time, but the effort is similar to that of trying to control a sneeze until the tension becomes so great that the tics 'explode'. Tics may increase in frequency and severity in exciting or stressful situations and decrease when a student is relaxed or engrossed in an activity.

Main characteristics

Physical tics

These range from simple blinking or nodding through more complex movements to more extreme conditions such as echopraxia (imitating actions seen) or copropraxia (repeatedly making obscene gestures).

Vocal tics

Vocal tics may be as simple as throat clearing or coughing but can progress to be as extreme as echolalia (the repetition of what was last heard) or coprolalia (the repetition of obscene words).

Tourette's syndrome itself causes no behavioural or educational problems, but other associated disorders such as ADHD or Obsessive Compulsive Disorder (OCD) may be present.

TOURETTE'S SYNDROME PROFILE

Potential impact of characteristics on learning	• Disruption of student's attention, performance and achievement. • Impact of tics on social relations and self-esteem. • Difficulties organising work, memorising and copying. • Repetitive behaviours. • Copying information quickly and accurately from the board can be particularly difficult.
Possible supportive teaching and learning strategies	• Accept the student; the tics are not deliberate. • Provide a tolerant, compassionate classroom ethos. • If the student is in agreement, discuss the impact of Tourette's syndrome with the class. • Provide a challenging but stress-free teaching and learning environment with the facilities to access a private area for study or when taking examinations if required. • Allow additional time for the student to complete work or examinations.

TOURETTE'S SYNDROME PROFILE

	• Allow the student to sit where he/she feels most comfortable.
	• Lower stress by giving instruction in stages.
	• Arrange a shared signal that the student can use to show that they need to leave the room as tics are becoming overwhelming, and a private place for them to go.
Preparation and planning for outdoor learning	• Explain expectations and tasks well in advance.
	• Provide friendship group working or 'buddy' system.

Further information

Tourettes Action
0300 777 8427 (Helpline)
www.tourettes-action.org.uk

4 Fieldwork and the outdoor classroom

> The environment is at the heart of everything we do and we believe being outside in different habitats offers exciting challenges that students may not otherwise get the opportunity to experience.
>
> The Field Studies Council (August 2017)
> www.geography.org.uk/resources/fieldwork/

Fieldwork is an essential component of geography education, enabling pupils to develop a better understanding of geographical features and issues in real life and extend their subject knowledge. In the best scenarios, however, fieldwork also provides opportunities for pupils to develop a range of social and communication skills that are difficult to develop solely in the classroom. Being outside can also aid motivation and independence and bring a completely new dimension to teaching and learning – features that are particularly valuable for pupils with special educational needs and/or disabilities (SEND). These pupils, for a variety of reasons, may have very limited experience of environments and cultures different to their own: it is still not uncommon to discover pupils who have not travelled to the nearest city or nature reserve, let alone other countries; some will not have strayed further than the housing estate where they live. Furthermore, simply being 'out of the classroom' can bring advantages to pupils who find academic work and 'sitting still' a daily challenge. Whether the immediate school grounds, the local area or locations further afield, a different environment with different types of activity can provide a boost to their interest and motivation, and ultimately to their achievement.

Well-planned fieldwork will enable pupils to:

- experience a range of environments
- increase their knowledge and understanding of different environments and develop their ability to identify geographical themes and features
- develop appreciation of, and respect for, the environment
- relate class work to the real world and their own observations
- investigate, question and hypothesise
- justify their thinking and conclusions
- collect, present and analyse data

- develop map-reading skills and the use of new technologies, e.g. geographical information systems (GIS)
- be creative and independent, fostering ownership of their learning and outcomes
- learn how to deal with uncertainty, keep safe and cope with acceptable levels of risk
- develop relationships with peers and teachers/teaching assistants (TAs) through shared experience: fieldwork often facilitates a more relaxed social contact among all participants

Working with students in the outdoor classroom is exciting, challenging and motivating (and often exhausting). As teachers, we experience hugely contrasting emotions, often questioning if the energy that we put into organising outdoor learning is worth the benefits. Fieldwork is, however, a requirement of the National Curriculum and is now compulsory for GCSE and A Level assessment. When well-planned and organised, fieldwork presents a wide range of learning opportunities for all pupils with SEND. (See Appendix 4.1.)

The National Curriculum states that fieldwork should be used to enable pupils to:

- observe, measure, record and present the human and physical features in a local area using a range of methods including sketch maps, plans and graphs and digital technologies. (KS2)
- collect, analyse and draw conclusions from geographical data, using multiple sources of increasingly complex information, in contrasting locations. (KS3)
- understand and apply specific geographical knowledge, understanding and skills to a particular and real out-of-classroom context. In undertaking fieldwork, students practise a range of skills, gain new geographical insights and begin to appreciate different perspectives on the world around them. (KS4)

(DfE, 2013)

Possible barriers to including pupils in fieldwork

Physical barriers are usually the most obvious issues to consider, especially when groups include one or more wheelchair users and/or pupils with mobility difficulties; these include:

- steps and steep gradients (especially in wet weather)
- uneven surfaces (not only in the countryside but also cobbled streets in towns and gravel paths/driveways)

- settings which are physically challenging and tiring (even for able-bodied pupils) such as lengthy hikes, mountainsides, quarries and multi-level buildings without lifts or escalators
- inaccessible transport
- a lack of specialist accommodation or facilities
- crowded locations such as busy city centres

**Less obvious barriers to successful participation
and enjoyment include:**

- difficulties with the legibility and/or complexity of information and task sheets, including diagrams and maps
- poor visibility/inadequately lit locations
- inaudibility of speakers/leaders in outside (windy) locations or inside noisy factories or amongst crowds
- lack of signers or induction loops for pupils with hearing impairment
- time-related issues such as unusually long (or unspecified) intervals between mealtimes, or limited opportunities to visit toilets
- lack of mains electricity can create difficulties for students reliant on micro-computer-driven equipment
- wide open spaces, cliffs or high buildings, underground caves and passages and the possibility of encountering particular forms of wildlife, which can also present challenges for phobic students

Pupils can practice using fieldwork equipment in the familiar surroundings of the school grounds

- crowded environments which may pose problems (e.g. for pupils on the autistic spectrum)
- a particular challenge when fieldwork includes questionnaires and face-to-face interaction with the public

There is also a need for careful consideration of appropriate toilet facilities, additional time required for loading and unloading wheelchairs from vehicles, and meal arrangements, such as meeting special dietary requirements and ensuring that pupils can cope with a self-service cafeteria.

For residential field courses, there are additional issues associated with domestic arrangements such as washing, sleeping and accessing social and educational facilities.

Planning fieldwork

It can happen that previously used locations for fieldwork, and established ways of working, have to be reconsidered in the interests of individuals with SEND. Teachers should plan well ahead and think about all the options; some questions to discuss with colleagues are suggested here:

- Why do we use this particular location? Are there alternatives which will ensure a more inclusive curriculum?
- Is there full information available about the destination at an early stage? Can we visit before we get too far along with the planning?
- What are the desired outcomes from this experience?
- Could the same outcomes be achieved through visiting more accessible locations or through different tasks or over a longer period?
- Can we employ strategies to make the destination and/or activities more accessible to students with SEND?
- What support and specific facilities will be required to ensure full access to the learning offered? Are there sufficient resources available? Who needs to be informed to ensure such support will be put in place? (Pupils with education, health and care plans may have personal budgets which can be used to contribute to additional costs.)
- Have the fieldwork opportunities been discussed with the pupil to establish specific mutual expectations?
- If, after careful consideration, the fieldwork appears to be inaccessible, what alternative experiences could be planned? Could the use of video, film or 'virtual reality fieldwork' replace an outdoor experience?
- Are there particular weather conditions that need to be considered (e.g. pollution impacts on students with asthma, sun effects on students with dermatological conditions)?

Planning activities

When designing fieldwork activities, we have to build in accessible learning opportunities and remove, or at least reduce, barriers to outdoor learning. This can involve consideration of a pupil's mobility and the physical access to fieldwork; the anxiety that may result when new locations and routines are involved; and the cognitive and self-organisation challenges that ensue. Medical considerations must also be taken into account of course, especially for longer and residential field trips; something that may not be an issue in an hour-long geography lesson becomes something significant to think about. Asthma, diabetes, epilepsy, musculoskeletal disorders, lung and kidney complaints, heart problems and chronic fatigue syndrome – all of these conditions are encountered regularly in mainstream schools and will usually (but not always) be documented. It is always important to check pupils' personal files and, where necessary, to speak with the students and their parents/carers about the most appropriate steps to be taken to enable them to access and enjoy outdoor learning. In many cases, an adult or responsible peer can be designated to keep an eye on an individual pupil; for example, if there is a diabetic pupil in the group, someone should be aware of what, when and how much they are eating; for other pupils, special medication may be required, and this will have to be administered by a designated member of the staff.

Group fieldwork first aid kit

4 Triangular bandages
1 Extra large, plain, sterile, unmedicated wound dressing
2 Large, plain, sterile, unmedicated wound dressings
3 Medium, plain, sterile, unmedicated wound dressings
2 Small, plain, sterile, unmedicated wound dressings
12 Assorted plasters [in packet]: Hypoallergenic
2 Sachets of sterile water/saline
10 Antiseptic wipes
4 Conforming [crepe] bandages at least 7.5 cm wide
1 Melolin dressing
1 Reel micropore adhesive tape (Non-stretch)
6 Assorted safety pins
1 Pair of blunt nosed, rustproof scissors
1 Pair disposable gloves
Glucose tablets [for diabetics]
Cling film for major burns/abdominal injuries
(The Royal Geographical Society, www.rgs.org/NR/rdonlyres/09D5E556-
   DF19-4A5B-9EA9-B079D64F006D/0/FW_Safety_Firstaidkit.pdf)

There is no standard template to support special needs students in the outdoor classroom; each pupil needs to be considered individually. Whilst students with special needs may experience certain challenges, they also have strengths, and it's useful to consider ways in which they can share these strengths in a group situation outdoors.

Before going on a field trip, or engaging in any outdoor learning, list the proposed activities and any barriers to learning that they might pose for your pupils. You may find Table 4.1 helpful in suggesting strategies to support pupils with SEND. All students will have a variety of needs, and the combination of these, together with their character, is what helps to make them unique.

Teresa Lenton, a secondary Post Graduate Certificate in Education (PGCE) course co-ordinator, constructed a useful mnemonic to help remind us of considerations when planning fieldwork experiences. You may wish to photocopy and annotate it at a departmental meeting. It will be useful to consider a piece of fieldwork that the department currently offers. Use this to annotate the mnemonic with any issues that reduce the quality of the experience for students with special needs that you have in an identified group. Now use Table 4.1 to consider how you would reduce these barriers to learning.

> **F**unding
> **I**CT?
> **E**qual opportunities
> **L**egislation
> **D**esign
> **W**here? When?
> **O**rganisation
> **R**isk assessment
> **K**inaesthetic learning

As with other aspects of the geography curriculum, learning objectives in fieldwork will be achieved at different levels of breadth and depth according to pupils' abilities: Table 4.2 signposts some possible steps of progression.

A note about risk assessment

Within your school you will have a member of staff who is designated as the Educational Visits Co-ordinator (EVO), who will work alongside you to advise on the school's, the local authority's and the DfE's guidelines for facilitating outdoor learning safely. The decisions that you take regarding reducing the barriers to

Table 4.1 Strategies to consider when planning fieldwork for pupils with SEND

Names of pupils:	
Teacher: TA:	
Date of field trip: Location:	
Strategy	*Details: who, when, how . . .*
Arrange a pre-visit to evaluate terrain accessibility, toilets and eating/accommodation facilities.	
Explain tasks before the visit; pre-read handouts to pupils.	
Use digital images/video to create a visual narrative about the place that can be used both pre- and post-visit.	
Create a sensory memory box.	
Use clear line sketches to emphasise information provided on maps and photos.	
Plan out a tactile/sensory trail, linked to the planned activities.	
Provide recording sheets – clearly linking the activity to its purpose. Consider coloured paper, larger print, simplified text and diagrams, writing frames and cloze procedure: hand-held recording devices. (A sample writing frame is included as Appendix 4.3)	
Consider preferred communicated system for outdoor learning. Avoid giving instructions in a location with a lot of background noise, or when standing in bright sunshine so that lip reading is difficult.	
Identify appropriate pairs/groups of pupils to support each other.	
Use additional visual information, e.g. photos, line drawings, symbols.	
Give a clear description of the sequence of events, explaining how the day is organised: make a visual timetable.	
Communicate during the visit what is happening and what will happen next, ensuring that sufficient time is being allowed to meet the needs of specific students, e.g. adequate opportunities for resting or meeting medical requirements.	
Provide clear and explicit guidelines, in an appropriate format, about expected behaviour.	
Plan for appropriate adult support, with clear guidance for TAs and other adults about their responsibilities and objectives.	
Identify and communicate appropriately potential dangers, and strategies that can be drawn upon to deal with them.	
Introduce the qualified first aiders (more than one) to everyone in the group, and ensure that an appropriate first aid kit is taken along. Any essential medicines should be in the care of a designated adult who has detailed information about their administration.	

Table 4.2 Progress in fieldwork

Shops and services enquiry progressing to comparison of two shopping centres in different locations, e.g. inner city area and commuter/edge of city area.

Suggestions for increasing the level of challenge in: types and amount of data, use of vocabulary, complexity of enquiry and development of higher order skills.

→ (increasing level of challenge)

Enquiry/map reading				
Responds to questions verbally or with signs/symbols, e.g. *What shops are there in the town?*	Uses four points of the compass. Follows simple enquiry, with (TA) support.	Uses eight points of the compass and simple grid references.	Uses town plans and maps to locate shops in and out of town.	What types of services does each location have? How does the site and situation of each location differ, what are the *demographics* of the local residents, what sort of services do they need, what is the *sphere of influence* of each location?
Understands locational and directional vocabulary (near/far, left/right, north, south).	Poses questions, e.g. *How well-used are the shops/services?* (How many customers in 30 minutes?).	Follows simple enquiry, e.g. *How far do people travel to shop?* Suggests reasons for findings.	*How well is each location catering for shoppers?*	
Recognises/names key features of the local town.	Vocab: adjacent, parallel, junction, location.	Vocab: accessible, convenient, settlement, sustainable.	*How could each one be improved?*	Vocab: threshold population, sphere of influence, leisure shopping.
			Vocab: high order/low order goods; supply and demand; market forces.	

Shops and services enquiry progressing to comparison of two shopping centres in different locations, e.g. inner city area and commuter/ edge of city area.

Suggestions for increasing the level of challenge in: types and amount of data, use of vocabulary, complexity of enquiry and development of higher order skills.

Data collection and recording →

Can collect data, e.g. the number and type of (shops) services in the High Street.	Identifies and categorises shops/ services, e.g. clothing, food outlets, supermarket household goods, chemist, hairdresser, café/restaurant, finance, tourism, fast food etc. Records some simple data (e.g. tally chart) and presents it in graphical representation, e.g. a bar chart with a simple scale using squared paper.	Different types of data (e.g. shopper survey, different types of housing, shops boarded up) collected, recorded and presented in a variety of ways, e.g. bar chart, pie chart.	Substantial data collected and recorded in a variety of ways, e.g. a bar chart, pie chart or graph. Use of software where appropriate.	Charts and graphs show differences between types and numbers of shops and services/shoppers, with written description, analysis and conclusions.
Identifies and categorises different types of shops/ services, e.g. clothes, food, household goods, chemist, bank, hairdresser, café/restaurant, finance, tourism, fast food etc.	Completes a cloze activity.	Short written description (writing frame provided).	Numbers of chain stores, independent shops, hairdressers, charity shops, eating places etc.	Some statistical analysis: mean, median and mode.
		Annotated sketches photo and simple colour-coded maps.	Transport links.	Written analysis and higher level, more advanced mapping technique – isolines.
			Cross referencing.	
			Written description, conclusions justified.	

Shops and services enquiry progressing to comparison of two shopping centres in different locations, e.g. inner city area and commuter/edge of city area.

Suggestions for increasing the level of challenge in: types and amount of data, use of vocabulary, complexity of enquiry and development of higher order skills.

Fieldwork activities and equipment

Looks at photographs of local town prior to visit. Visits town to establish number and types of services/shops. Makes a simple plan of the High Street/town centre (perhaps using pre-prepared images/symbols to glue on – or using a key of letters/colours).	Makes sketch/plan of town centre with main stores/features. Or annotates enlarged base map (Gode). Uses digital camera, click counter.	Designs questionnaire for shoppers to record frequency of visits, mode of travel, distance covered, reason for visit (may also add availability of parking, reasons for using other service centres). Verbal survey conducted. Hand-held recording device. Collation sheet.	Sampling techniques used and reasons given. Consideration of people's responses; inferences made and data interpreted to reach informed conclusions.	Visit to town planning offices to hear plans for development: question-and-answer session. Critical reflection. Pupils reflect on information gathered and suggest how to develop or improve shopping in each location. Debate, e.g. 'High street vs retail park'; this can be filmed (for assessment). Design a new shopping location. Annotated sketches and plans. Pupils reflect on fieldwork study and evaluate: what worked/didn't work? How could it be improved?

learning should form part of your risk assessment. The Field Studies Council provides valuable information about the content of risk assessments: www.field-studies-council.org/ (see other sources of information at the end of this chapter).

The risk assessment can be a hugely valuable document for all staff and students if it is a 'live' document. You will find it useful to share this with other teachers and TA colleagues who accompany the visit. It may also be appropriate to use some or all of it, either in its existing or in an adapted format, with your students, parents and carers. Pupils themselves may be able to contribute to the risk assessment and suggest appropriate strategies for minimising risk – especially if shown visual/video footage. This will help to clarify expectations and make everyone aware of the strategies to use should the need arise. The more informed fieldwork staff and participants are, the more likely it is that they will have a successful, enjoyable and safe experience. (See Appendix 4.2 for a sample risk assessment template.)

A checklist for support

- Give out all fieldwork details in advance in appropriate formats, e.g. spoken, written, symbols, pictures, so that queries can be sorted out prior to the visit. Provide details of activities to be undertaken – with learning objectives. Explain potential emergency situations and how pupils will be expected to behave.
- Provide clear guidance on what to wear, any special equipment needed, water to carry, sunscreen, sandwiches/snacks etc. It can be useful for the leaders to wear something brightly coloured so that they are easily and quickly identified when needed.
- Summarise for hearing-impaired students any debates/role play activities. Ask all students to raise their hand before they speak so that lip readers can face them in a debate situation.
- Travel – are there any particular issues related to getting to and from a particular site? What about access at the site itself?
- Link with a local university department which may have access to inject facilities that create tactile maps. These allow visually impaired students to navigate maps by touch.
- If the visit involves 'look and see' type activities consider using carefully selected digital images to supplement the experience. These can be accompanied by clear line sketches.
- Extra time can be of benefit for all. Do not rush – the temptation to overload the day can reduce rather than enhance learning.
- Have an appropriate communication system in place, with all adults able to contact the field course leader and each other, as well as a designated colleague back at school (in case of emergencies). Remember that mobile phone signal strength may be weak or non-existent at the location you are visiting.

- If you are able to have TAs with you on the visit, make certain that they have been appropriately briefed about what to do, when and with whom.
- When completing questionnaires, think about using a buddy system or facilitating the use of a hand-held recorder.
- Consider distractability in the field. Try not to give instructions in places with lots of background noise, or with lots of activity going on, or standing with your face in shadow (this impedes lip reading).
- In planning for fieldwork that includes wheelchair users consider:

  - time: it may take longer for the students to get on and off vehicles, or in and out of particular locations
  - access to toilets
  - adequate breaks so that students don't become uncomfortable and can change position
  - weather: wheelchair users may need to consider more or different clothes
  - carrying clipboards on a wheelchair needs thinking about!
  - specialist transport may have financial implications; try to make certain that you are aware of this
  - who is to push the wheelchair?

Many departments find that the time invested in creating digital, video, audio and sensory resources related to favourite fieldwork localities is of huge value to many students. They take time to construct, but over time the benefits will be there for all to experience. Some departments create 'place boxes' containing photos and video of landscapes, simplified maps and samples of soil, sand, shells and pebbles. Symbol-supported resources can also be valuable with pupils who have more significant needs (www.widgit.com/resources/curriculum/geography).

Other departments have collated their various locality resources and created accessible websites that students can use both before and after the visit itself. Using a computer or tablet with digital images and information pages, before and after the visit, may help the students (including those on the autistic spectrum) to access the field learning in a more secure and comfortable way. They may feel more in control of their learning.

If you are planning to create a web-based or visual resource to support pre- and post-visit activities here are a few things to consider. You will need to think about how you use colour to convey meaning. This can unwittingly create barriers to learning for colour-blind users. A sharp contrast between background and text can be disorientating and uncomfortable for dyslexic users, whereas insufficient contrasts can make access difficult for those with visual impairments. One solution could be to make the design flexible, so that it is easy to change the colours and text size by making quite simple adjustments. (See Chapter 2 for further suggestions.)

Phoebe is a dyslexic student going on a day's fieldwork in the Peak District. In order for Phoebe to access the learning, her teacher and TA made certain of the following:

- Travel arrangements and timings were made clear in advance, with reminders given the previous day.
- All fieldwork information and handouts were provided in a variety of formats, verbal, written (on coloured paper) and emailed to her home. She was given a list of key words and place names.
- Her TA spent some time in lessons working with her on time management and organisational skills.
- Phoebe was allocated to work with a group of supportive peers who helped with reading out information during the fieldwork.
- She was allocated some TA support during one lunchtime to complete her recording of the information collected and prepare for her presentation.

Using the school grounds to support learning outdoors

It can be helpful for students to practise and develop fieldwork skills in the familiar environment of the school grounds. Outlined here are several ways that the grounds can be used both as part of the geography curriculum and as preparation for fieldwork visits off-site:

- using a compass, anemometer and clinometer
- developing trails
- treasure hunts/orienteering
- using our senses to investigate
- following directions
- investigating weather (where are the hottest places, the shadiest places etc.)
- decision-making (where should we put litter bins, noticeboards, benches)
- expressing likes and dislikes
- placing images/representing places
- measuring rainfall

- creating questions
- developing geographical stories (What if . . . ?)
- developing an awareness of change

Resources

The Geographical Association has a variety of resources, including the 'Fieldwork File' poster www.geography.org.uk/download/REfieldworkposter.pdf

The Royal Geographical Society RGS-IBG website (www.rgs.org) provides information about the Expedition Advisory Centre, which offers information, training and advice for anyone embarking on scientific or adventurous expeditions.

AQA provides a variety of resources for all aspects of geography teaching www.aqa.org.uk/resources/geography/gcse/geography/teach/fieldwork-toolkit

For fieldwork ideas and resources, go to http://geography.org.uk/resources/fieldwork/ideasandresources

Fieldfare works with people with disabilities and countryside managers to improve access to the countryside for everyone www.fieldfare.org.uk

The 'Phototrails' website allows users to view countryside routes through a series of photographs and descriptions of the path features, taking in surface, path width, gradients, barriers and amenities such as seats, disabled parking and accessible toilets www.phototrails.org

5 Monitoring and assessment

Assessment matters

> Somewhere in the evolution of education, assessment as a means for enhancing learning has become distorted to become assessment for bureaucratic, recording and accountability reasons.
>
> Spendlove, 2009

What is assessment, and why does it matter?

Assessment fulfils many purposes; within this chapter assessment will be considered as a means of enhancing learning for pupils with SEND. Table 5.1 contains some varied definitions of assessment. Reflection on them can be a useful individual task, but first we highlight some key general principles of assessment.

- It is an ongoing process which should enhance learning.
- It has a variety of purposes.
- It exists in a variety of different forms. 'Using one assessment for a multitude of purposes is like using a hammer for everything from brain-surgery to pile-driving' (Haney, 1991).
- It is a process involving pupil feedback.
- It is a means of measuring not just what pupils know but also what they understand, how they can apply knowledge and what skills they have.
- It carries an impact that should be carefully considered.
- It is accurate and successful when it has been carefully crafted by teachers who understand what assessment is about.
- It yields an outcome and provokes a response.

It is important to remember that 'assessment' is not synonymous with 'measurement'. It will sometimes include aspects of measurement in terms of scores, grades and results; however, there is a whole range of techniques that teachers use every day to monitor how well pupils are doing – e.g. questioning, observation, completion of tasks – and these are equally relevant and useful.

Table 5.1 Definitions of assessment

Assessment is inevitably part of every teaching activity. "How is she getting on?" "Did he understand that?"

Assessment is therefore too important to be the sole preserve of national tests and assessments. (Report of the NAHT Commission on Assessment, February 2014)

'A tool or method of obtaining information.' (National Council on Measurement in Education)

'Assessment is very important to tracking progress, planning next steps, reporting and involving parents, children and young people in learning.' (Education Scotland)

'Assessment is the process of gathering, interpreting, recording and using information about pupil's responses to educational tasks.' (Lambert, 2001)

'The systematic collection, interpretation and use of information about learning'. (Council for the Curriculum, Examinations and Assessment)

Assessment is a key professional competence which enables teachers to make teaching and learning more effective. It is an important part of motivating students and has considerable impact on their learning, engagement and self-esteem, particularly for those who lack confidence due to learning difficulties.

Assessment requirements

The DfE Teachers' Standards of 2011 make several statements regarding monitoring and assessment. Teachers should:

* be accountable for pupils' attainment, progress and outcomes
* be aware of pupils' capabilities and their prior knowledge, and plan teaching to build on these
* guide pupils to reflect on the progress they have made and their emerging needs
* know and understand how to assess the relevant subject and curriculum areas, including statutory assessment requirements
* make use of formative and summative assessment to secure pupils' progress
* use relevant data to monitor progress, set targets and plan subsequent lessons
* give pupils regular feedback, both orally and through accurate marking, and encourage pupils to respond to the feedback

These guidelines are comprehensive, covering recognition of prior learning, monitoring, assessment of various types and, importantly, pupil reflection on progress. All issues will be touched on within this chapter, but firstly we provide a summary of assessment requirements in geography across the age groups, including options for those with special educational needs. It is worth mentioning that the style of assessment needs to be carefully considered when departments select qualifications; the aim is to support learners and enable them to show what they can do rather than what they can't do.

The National Curriculum (ages 5 to 14)

The National Curriculum (NC) is a statutory programme of study (for all local authority maintained schools) with attainment targets to cover Key Stages 1–3, ages 5 to 14. The NC states that 'by the end of each key stage, pupils are expected to know, apply and understand the matters, skills and processes specified in the relevant programme of study'. This summative statement illustrates the key elements of study in geography:

- factual knowledge
- understanding of concepts and processes
- use and application of skills
- involvement in practical fieldwork

The aims set out at the start of each key stage in the programme of study provide a foundation for expectations of attainment by the end of each key stage, but beyond this, assessment becomes an issue for the individual school. Level descriptors for the assessment of key stages have been removed. This leaves schools and geography departments with the opportunity (and challenge) to devise their own assessment strategies for progress and attainment at each key stage.

GCSE (ages 14–16)

New GCSE geography specifications were introduced to be taught from September 2016. A new numerical grading system of 9–1 replaces the letter grades. Content is set out in detailed criteria, and the inclusion of fieldwork has been strengthened. The changes to assessment will present a challenge to many learners who find extended writing difficult and may be unsettled by reading questions which seem unattainable to them. These pupils can no longer benefit from modular assessments and project-based controlled assessment, and examinations and revision of large quantities of material will be particularly daunting for them. Changes to GCSE assessment include:

- Untiered exam papers.
- Increased expectations for extended writing in examinations.
- Controlled assessment and modular examinations have been dropped.
- All examinations are terminal.

Sample Assessment Materials (SAMs) are produced by all examination boards (AQA, Edexcel, Eduqas and OCR). These materials should be given full consideration when departments select a specification as the 'style' of assessment varies. The following assessment objectives are common to all specifications:

> AO1 Demonstrate knowledge of locations, places, processes, environments and different scales.

> AO2 Demonstrate geographical understanding of different concepts and how they are used in relation to places, environments and processes and the interrelationships between places, environments and processes.

> AO3 Apply knowledge and understanding to interpret, analyse and evaluate geographical information and issues and make judgements.

> AO4 Select, adapt and use a variety of skills and techniques to investigate questions and issues and communicate findings.

In each specification as a whole, 5% of the marks are used to credit the accuracy of spelling, punctuation and grammar and the use of specialist terminology.

A Level (ages 16–18)

New A Level specifications were introduced in geography for first teaching in 2016. Changes include the following:

- AS and A Level are now stand-alone qualifications.
- Fieldwork is compulsory and examined.
- There is a teacher-assessed independent investigation at A Level.
- Both AS and A Level have a terminal (not modular) exam structure.

Sample Assessment Materials (SAMS) have been published by all examination boards. The following are the common assessment objectives although the style of assessment varies across the board.

> AO1 Demonstrate knowledge and understanding of places, environments, concepts, processes, interactions and change at a variety of scales.

> AO2 Apply knowledge and understanding in different contexts to interpret, analyse and evaluate geographical information and issues.

> AO3 Use a variety of quantitative, qualitative and fieldwork skills to investigate geographical questions and issues, interpret, analyse and evaluate data and evidence; construct arguments and draw conclusions.

It must be remembered that A Level presents challenges for differentiated teaching and learning just as other levels of education do; there can be wide disparities between ability levels within an A Level teaching group. There must be careful planning for meeting individual needs given the more challenging content and assessment styles.

Entry Level Certificates

Two examination boards (OCR and Edexcel Pearson) offer Entry Level Certificates to learners who may not be able to access GCSE level. The qualifications have been adapted so that they can be co-taught with the new 2016 GCSEs and are 100% internally assessed through tasks, e.g. a personal project presented in any appropriate manner (PowerPoint, written format, display) and tests (which can be retaken) that can be taken at any point during the course.

Unit Award Scheme

The AQA Unit Award Scheme (UAS) offers certificated achievement to learners who cannot meet the challenge of GCSE. A wide-ranging and flexible scheme, it can be used to engage learners and raise the self-esteem of pupils struggling academically. There is no limit on what can be accredited (enabling co-teaching with other qualifications), how long it takes or how learning is evidenced.

ASDAN

The ASDAN Geography short course has been authored by the Geographical Association and can be offered to all learners in a variety of settings and time periods. The focus is on completing challenges and skills based on individual ability rather than assessment against specific levels. The qualification is internally moderated and progression is offered to AoPE (Award of Personal Effectiveness) and Wider Key Skills.

By Key Stage 4 pupils will start working towards national qualifications, with some working towards certificated schemes as already outlined. Whatever the choice, pupils' achievements can be recognised and acknowledged through qualifications and certificates that are matched to their specific learning needs.

Types of assessment

An effective assessment system should make use of as many different types as possible; Table 5.2 provides a summary of the main types of assessment.

Table 5.2 Different types of assessment

Assessment types	Definition	Comment
Formal assessment	Standardised measure with supporting data for the conclusions made. Includes tests and examinations.	These types of assessment are usually summative and cover external examinations.
Informal assessment	Assessment information collected under 'normal classroom conditions', usually over a period of time.	These assessment techniques include a wide range of 'everyday' classroom practices, e.g. questioning, completion of tasks.
Formative	A range of formal and informal assessment procedures used by teachers during the learning process so that they can modify teaching and learning activities to improve pupil attainment. (Council for the Curriculum, Examinations and Assessment (CEA))	As a subject, geography is well placed for the use of a range of formative assessment types – discussion, presentations, data skills, practical skills, fieldwork, projects, explanation through diagrams, spatial skills and written work.
Summative	Comes at the end of a learning sequence and is used to acknowledge, record and report on pupils' overall achievement at a given point. (CEA)	Summative assessment can be internal (school based) or external (national examinations). It is often used to plan the next step and to provide feedback to parents and pupils after an extended period of learning.
Diagnostic assessment	Used to identify individual strengths and areas for improvement and to inform next steps. (CEA)	Diagnostic assessment can also be used to find out what pupils already know, understand and can do.
Evaluative assessment	Concerned with the overarching performance of arrangements in a department, school or system. (CEA)	Assessment exists at a variety of scales – from the whole school to a subject department to an individual teacher; together they must form a cohesive system.
Subjective assessment	Assessment based on personal feelings and opinion; has more than one correct answer.	Subjective assessment has relevance in geography, for example for topics in human geography regarding personal views and perceptions of place or empathising with individual opinions on issues such as migration and climate change.

Assessment types	Definition	Comment
Objective assessment	Clear right and wrong responses, judged on criteria which have a high degree of reliability.	Uses in geography relate to quantitative skills such as those used in statistical analysis of fieldwork.
Descriptor-based assessment	Assessment according to a set of descriptors which can be used in a formative or summative way.	Level descriptors have been used in the National Curriculum in the past and are commonly used by examination boards in mark schemes.
Exam-based assessment	Isolated, formal assessments which are summative in nature. Exams are taken under formal and standardised conditions.	Exams are included in GCSE and A Level Geography assessment. In the renewed specifications (for teaching in 2016) they include terminal rather than modular exams. A range of support exists for pupils with SEND; this is discussed at the end of this chapter.

Assessment for learning

Teaching and assessment are inextricably linked. Assessment for Learning (AfL) was an initiative formulated in the late 1990s with the aim of using assessment to help pupils learn. It offered an alternative to the traditional, narrow, mainly summative styles of assessment that tended to encourage 'teaching to a test' and did not suit all learners. AfL offered an approach which prioritised learning and understanding with ongoing, built in assessments which redressed the balance of formative and summative assessment. 'AfL is the process of seeking and interpreting evidence for use by learners and their teachers to decide where the learners are in their learning, where they need to go and how to get there' (Assessment Reform Group, 2002).

AfL is worthy of some consideration here as it offers the flexible, pupil-centred approach that is needed when considering assessment for pupils with SEND, and it continues to be widely adopted at a personal and whole-school level as a philosophy for learning and assessment.

Key principles of assessment for learning

Assessment for learning should

- **be part of effective planning of teaching and learning**

  A teacher's planning should provide opportunities for both learner and teacher to obtain and use information about progress towards learning goals.

Assessment processes are an essential part of everyday classroom practice (questioning, completion of tasks) and involve both teachers and learners in reflection (observation, interpretation and judgement), dialogue and decision-making.

- **be regarded as a key professional skill for teachers**

  Teachers require the professional knowledge and skills to plan for assessment; observe learning; analyse and interpret evidence of learning; give feedback to learners and support learners in self-assessment. Teachers should be supported in developing these skills through initial and continuing professional development.

- **be sensitive and constructive because any assessment has an emotional impact**

  Teachers should be aware of the impact that comments, marks and grades can have on learners' confidence and should be as constructive as possible in the feedback that they give.

- **take account of the importance of learner motivation**

  Assessment that encourages learning fosters motivation by emphasising progress and achievement rather than failure. This is particularly so for those pupils with SEND; comparison with others who have been more successful is unlikely to motivate and encourage learners. It can also lead to their withdrawing from the learning process. (See Table 5.6 at the end of this chapter.)

- **promote a shared understanding of the criteria by which they are assessed**

  For effective learning to take place learners need to understand what it is they are trying to achieve – and want to achieve it.

- **enable learners to receive constructive guidance about how to improve**

  Learners need information and guidance about how to improve. This may involve a teacher highlighting strengths and giving advice on how to develop them and also being clear and constructive about any weaknesses and how they might be addressed.

- **recognise the full range of achievements of all learners**

  Assessment for learning should be used to enhance all learners' opportunities to learn and enable all learners to achieve their best and have their efforts recognised.

(See Appendix 5.1 for a list of activities for assessment.)

Putting assessment into practice

Assessment activities

The main consideration when designing assessment activities which are inclusive of pupils is to use a range of strategies which provide frequent opportunities for feedback. Geography as a subject is well placed to draw on a range of assessment tasks, individual or based on pair or group work. In addition to writing skills, creative, design, numeracy, presentation and information technology skills can also become a foundation for the assessment of geographical knowledge and understanding. (See e-resources for an example of assessment through an annotated 'earthquake poster'.) The following examples of assessment activities address two of the changes to formal assessment at GCSE: increased use of numeracy and extended writing.

Quantitative skills are used in geography to analyse data which supports theories and concepts and to explain fieldwork results. A stepped assessment task (Table 5.3) can aid explanation and build confidence in mathematical skills, moving from an example to a partly completed question and finally a fully independent activity. Many of the new examinations at GCSE require extended writing, and an activity to show pupils what is meant by 'extended writing' is shown in Table 5.4.

Table 5.3 A stepped activity for assessing numeracy in geography

---

Calculating an average or mean

After providing a basic explanation and a short definition, a stepped task can be used to differentiate assessment.

Example

Immigration to UK, 2005–2014 (thousands). Source: Office for National Statistics (ONS, 2017).

2005	2006	2007	2008	2009	2010	2011	2012	2013	2014
567	596	574	590	567	591	566	498	526	632

Calculate the average value for this data set

Calculation.

Step 1: Add the number of migrants each year to find the total value for the ten year period.

$567 + 596 + 574 + 590 + 567 + 591 + 566 + 498 + 526 + 632 = 5707$ thousand

Step 2: Divide the total by the number of years, in this case 10.

$5707 \div 10 = 570.7$

Partly completed question

The following rainfall totals (mm) per month were recorded in Oban and Norwich in 2014.

	J	F	M	A	M	J	J	A	S	O	N	D
Oban	150	75	98	68	76	89	55	80	98	130	128	110
Norwich	42	39	44	39	36	52	45	51	63	59	65	61

1   (a) **Calculate the average monthly rainfall for Oban and Norwich.**

Step 1: Calculate the total annual rainfall. For Oban:

150 + 75 + 98 + 68 + 76 + 89 + 55 + 80 + 98 + 130 + 128 + 110 = 1157 mm

Step 2: Divide this by the number of months.

1157 ÷ 12 = _____ mm

Complete the same calculation for Norwich:

Step 1: Calculate the total.

42 + 39 + 44 + 39 + 36 + 52 + 45 + 51 + 63 + 59 + 65 + 61 = _____ mm

Step 2: Divide this number by the number of months.

Total mm of rain ÷ number of months = _____ mm
Analysis

(b) **How many months had higher than average rainfall in Oban?**

Independent activity

Number of vehicles per hour in the town centre.

8am	9am	10am	11am	12pm	1pm	2pm	3pm	4pm	5pm
125	120	80	70	76	85	58	70	100	120

1   (a) Calculate the average hourly traffic during the day.
    (b) At what times of the day is traffic above the hourly average?
    (c) Account for the peaks in traffic at 8am and 4pm.
    (d) Why is 11am the quietest time of the day?

Based on one of the activities in *Maths in Geography*, with kind permission of Hodder Education.

Further considerations for assessment tasks include:

- Set realistic time targets for the completion of tasks – many pupils with SEND have a poor concept of time and may need longer than their peers.
- Make sure that deadlines are recorded, and set interim deadlines in an extended task.

- Work with the SENCO and teaching assistants so that any curriculum support time outside of class can be used effectively regarding assessment tasks.
- Consult Individual Education Plans (IEPs) to identify priorities for learning. These will often have a literacy or numeracy focus, and/or targets for improved behaviour, but can usually be applied in some way to geography subject matter, enabling the teacher to contribute to the pupil's personal plan.
- Present tests and assessments in oral instructions so that the assessment is not influenced by a lack of reading ability.

Always take time to read through assessment tasks when they are handed out and give the opportunity for clarification from pupils.

Table 5.4 Building extended writing

Example GCSE Question. Evaluate a named coastal management scheme. (6)

Step 1. Understanding the question. (This could be annotated on a whiteboard with the help of the pupils.)

Evaluate (How effective is the scheme? Give a balance of positive and negatives.) a (just one needed) named coastal management scheme (measures to protect a coastline).

Step 2. Understanding the assessment.

Assessment objective 3 (AO3) Can evaluate geographical issues.

Marking levels: 5–6 (Well-developed ideas) 3–4 (Developed ideas) 1–2 (Simple ideas)

Step 3. Building a simple, developed and well-developed answer. (Simple ideas are written first, and developed and then well-developed content shown in brackets. These could be colour-coded, or the answer could be projected onto a whiteboard and pupils could help identify the statements that make the answer simple, developed and well-developed.)

Walton-on-the-Naze in Essex suffers from coastal erosion (due to the type of rock) (due to soft clays called London Clay). Coastline erosion is severe (due to erosion and cliff collapse) (and powerful longshore drift which removes sand that would protect the cliffs from erosion).

A management plan was put in place (in 1977) consisting of a sea wall (breakwaters and groynes) (and also the cliff was regraded – creating a more gentle slope).

The sea wall was expensive but very effective at protecting the residents. (The groynes stopped longshore drift removing the beach; this was low cost, and the trapped sand helped protect the cliffs.) (Because the groynes were not completely effective, breakwaters – larger groynes – were also needed; this increased the cost of the project. The regraded cliff was planted with trees and shrubs to stabilise it. This is low cost and environmentally friendly as it looks more natural.)

The project was not successful (the coastline is still being eroded due to the powerful longshore drift) (between 1999 and 2003, 300 tonnes of additional material was removed).

Questioning

Questioning is a key element of verbal and written assessment. Socrates defined teaching as 'the art of asking questions'. Socrates was no doubt referring to high order questions; however, questioning (in its various forms) remains one of the most commonly used teaching skills, second only to giving explanations. Verbal questioning for learning is instinctive; there is often little time allowed for planning, and few questions require much thinking from learners. Table 5.5 looks at some practical suggestions to make the most of questioning as a form of assessment for learning in geography.

Assessing, recording and feedback

Assessing

Teachers need to use a range of assessment strategies including:

- discussion (whole group, in groups, in pairs)
- observation of pupils
- marking and grading work
- providing written feedback
- tests
- individual assessment
- peer assessment
- portfolio

Marking and feedback are the most commonly used assessment strategies. In order to generate accurate and useful assessment, time must be invested in explaining the assessment criteria to pupils so that they are clear about what is required. The teacher should also give detailed and specific feedback (see the section 'Feedback').

Pupil and peer assessments are extremely useful strategies to enable pupils to fully understand and benefit from assessment. This may be achieved through pupil participation in setting assessment criteria and through self- and/or peer assessment. However, it is not easy. Pupils find self- and peer assessment difficult; there are feelings of embarrassment and uneasiness, which means that pupils are unwilling to commit to a decision. They don't want to offend or demoralise their peers, and they don't want to appear over-confident in their own self-assessment or admit to shortcomings. For these reasons, it may take time for pupils to become comfortable with, and competent in, procedures of self- and peer assessment. I used an assessment cover sheet for essays with a class of A Level students which included the essay title, final mark, detailed feedback comments and a short bullet point focus for next time; there was also an opportunity for the student to complete a section of self-assessment

Table 5.5 Questioning in geography

Planning questions

Pre-planning of questions is rarely part of lesson planning; questioning is commonly a spontaneous act. However, the right questions enhance learning and are an important part of ongoing assessment. Teachers can enhance questioning by:

- Pre-planning two to three key questions within the lesson.
- Before the lesson ask pupils to pre-plan a question.
- Using internal lesson observations for appraisal as an opportunity to focus on questioning technique. This can include a simple grid within the lesson observation notes to record the frequency of use of different types of questions, e.g. open, closed, random, targeted, questions based on factual recall, explanation, evaluation, wait time.
- Planning learning objectives as key questions to be answered.
- Many specifications are structured in 'key questions'; plan sub-questions as a pathway to learning.

Balancing the types of question asked

Many teacher questions are procedural – have you finished? where is your homework? Try and become aware of the type of questions you ask – are they always closed? How often do you ask higher order questions that require explanation or opinion rather than a single word? Do you ask more open than closed questions? Are your questions addressing different learners? Do your questions always relate to recall of facts? Or start with the same stem?

Remember higher level questioning

Examples of higher level questioning in geography topics include:

- Those that require explanation:

How is relief rainfall formed? Why does increased wealth lead to higher energy consumption?

- Those that require analysis:

What are the benefits of landfill sites over incineration of waste? Why do some governments want to limit immigration?

- Those that relate to decision-making:

What is the best way to reduce flooding in . . . ? What is the most effective way to reduce traffic in urban areas?

- Those that require evaluation:

To what extent is wind power a factor in coastal erosion? Are there more advantages than disadvantages of living in a city?

- Those that require application of knowledge:

Where would be the best location for a sports centre in . . . ? Why are there more beaches on the coast of . . . than the coastline of . . . ?

Logistics

Classroom etiquette – it is important to create a respectful listening environment for questioning. Dialogue is essential in a classroom, and the teacher should shape the functioning of classroom interaction that is inclusive and respectful – consider the following in terms of what you want to achieve:

- No shouting out
- Hands up policy
- Vary the time you wait for an answer; give pupils time to think. It can be useful to give additional time to pupils with SEND: 'Here's a question for you, Thomas . . . Have a think and I'll come back to you in a few minutes.'
- Don't focus on those who are always involved (either because they are very able or because they are disruptive); involve the whole class whilst keeping in mind that involving those reluctant to put themselves forward requires sensitivity.
- To involve reluctant pupils, consider placing a question that you know they can answer easily, use whiteboards to be held up with an answer or ask pupils to prepare answers to higher level questions in pairs. Don't always ask questions that have a right and wrong answer; ask questions which require an opinion and where there is no right or wrong, which reduces risk.
- Questions can be used to add pace and discipline, or to regain pupil focus.

Handling difficult questions, dealing with responses that are incorrect

A positive experience in classroom contributions can boost the self-esteem of pupils with SEN; therefore, it is essential that incorrect responses are carefully handled.

'Not quite . . . You're almost there . . . Keep thinking . . . I like the way you're thinking/having a go.'

Sometimes a pupil will understand but be unable to accurately articulate an appropriate response. When possible, offer a good model for them: 'I think this is what you're saying'; 'Let's see if Jason agrees . . . This is what Ellie is thinking . . .'.

Pupils' questioning

Pupils may also need encouragement to ask questions particularly if they have SEN and feel that the questions they want to ask will draw attention to themselves. It is extremely difficult for most pupils to put forward questions that seek to clarify understanding or further their understanding: they do not want to expose a potential academic weakness, they feel embarrassed, they feel uncomfortable in apparently 'questioning' the teacher. The reality is that if one pupil is unsure, there will be others. To gain pupil confidence in asking questions:

- set up the interactive whiteboard for pupils to 'post' questions at the end of the lesson
- ask for any concerns or progression questions to be written down and handed in at the end of the lesson
- ask pupils to come up with any concepts they didn't understand within a short group activity at the end of the lesson.

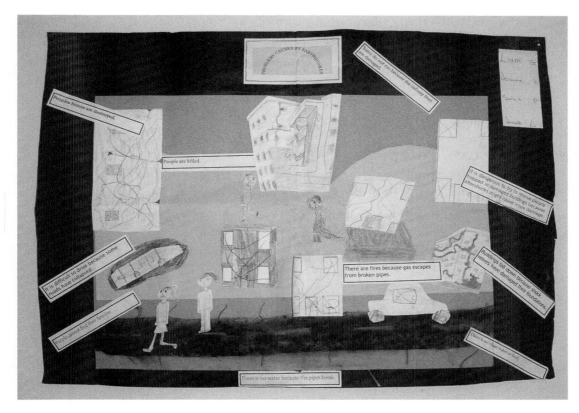

A student identified some of the effects of earthquakes on people and places and worked in a group to produce this display

in which they entered a predicted grade for the piece of work – it took time for pupils to have the confidence to complete this.

'Progress' by pupils with SEND can be evidenced in many ways. Assessment of knowledge, skills and understanding will dominate, but the following should also be included and can be equally powerful in boosting self-esteem and confidence:

- participation in class, including verbal participation
- involvement in group work and pair work – both of which can be uncomfortable for SEND pupils
- increased autonomy in tasks
- reduced dependence on support staff

Recording

Assessment records can be kept in various ways including:

Weekly records

- marking (within the whole-school and departmental policy)
- annotated pieces of work – examples can be kept as a departmental resource to help with standardised marking

- teacher notes – particularly useful during fieldwork
- in-depth formative comments on pupils' work (see the section 'Feedback')
- comments on IEPs
- video/audio/photographic evidence

Termly records

- class records (these can be stored as electronic mark books so that a Head of Department can have access and review the attainment of the 'geography cohort')
- pre-and post-unit assessment marks
- student awards
- progress against target judgements

Yearly records

- statutory review of education, health and care plans
- progress files
- moderated evidence, e.g. external non-exam assessments
- records of external assessments

Not all pupils will make steady progress, and often progress is stepped rather than advancing as a continual upward trajectory. Teachers must recognise that, because of their learning difficulties, some pupils may reach a plateau (temporary or permanent) in their achievements, or even regress. As a result, it may be necessary to revise targets and predictions, and it is therefore important to keep full and accurate records to support such decisions.

Feedback

Assessment provides a judgement on current attainment and should also indicate where improvement can be made. Teachers need to communicate both of these aspects in a clear and objective manner. With advancements in assessment procedures over the past decades, it is now widely accepted that '7/10 good effort' is feedback that falls widely short of good practice. Why 7 and not 8? What assessment criteria does the 10 mark comprise? What does 'good' effort mean? What do I need to do to get 8+? Feedback on formative assessment is part of a 'process':

- Schemes of work have built in assessments.
- As part of lesson planning the teacher decides what formative assessment is needed.
- Teacher sets the assessment task (keeping in mind variety, differentiation, time limit).

- Teacher introduces the assessment task in the lesson and spends some time reading through it, explaining what pupils need to do and allowing time/opportunity for clarification from pupils.
- Pupils complete the task.
- Teacher marks the work and gives written feedback.
- In class, the teacher makes some general comments about what was done well and not so well.
- Pupils are given time to reflect on the assessment – this may be individually or in groups and may involve pupils feeding back to the teacher on difficulties they had.

Teachers also need to be aware that comments used in assessment feedback need to be specific. Phrases such as 'more depth and detail needed', 'too descriptive', 'more analysis needed' and 'your explanation is difficult to follow' may seem straightforward to a professional practitioner, but may leave a pupil frustrated and struggling to understand what the teacher wants. The assessment activity in Table 5.4 illustrates an assessment task that can be used to show pupils what 'description' and 'explanation' look like and can lead to more 'depth'. (See Appendices 5.2–5.4 also.)

External examinations

Preparing for formal assessment in geography

All pupils find examinations and examination preparation difficult and stressful – the task is even more challenging for SEND pupils. Particular challenges include:

- feeling anxious
- lacking the motivation to sit an exam
- finding it difficult to focus during the preparation time and during the exam itself
- having problems with literal understanding of questions
- having difficulties processing information
- feeling anxiety about time pressures
- finding it difficult to cope with changes to the 'normal' routine of the day

The following provides some practical advice on how to manage the challenge of external examinations.

Planning

Planning, preparation and organisation can relieve some of the pressure of examinations and relate to more than just drawing up a revision timetable.

It may help pupils unsettled by the unfamiliar situation, if time is taken to show them where the examinations will take place, to explain the procedure and the formal instructions that will be given out and to explain the exam 'conditions' and the rules that need to be adhered to. Pupils with SEND are able to apply for special considerations as outlined below. If they receive support, then it is important to allow pupils to use these considerations in an informal setting, i.e. an internal test or exam prior to the 'real' event. Ideally the staff acting as a reader or scribe should be someone already familiar to the pupil.

It is helpful to consider individual requirements which help prepare for exams: when is the ideal time to study, what is the best environment to study in, what is the most appropriate length of study time, whether it is possible/helpful to have a revision 'buddy' to work with, what revision strategies work best (examples are outlined under 'Special Consideration' below).

A well-thought-out revision timetable is essential. It may need to take into account preparation for several subjects, and for each subject, a range of topics needs to be subdivided into manageable and realistic periods of time. Once set up, the timetable needs to be monitored and re-evaluated as necessary. It is essential to be realistic and build in rest and relaxation time, and care must be taken not to make the overall task look overwhelming and unmanageable; facing the scale of the task is something which in itself creates anxiety, i.e. coming to the conclusion that time is short.

Revision notes

Revision notes can take a range of formats, and it is up to the individual to select those that work best for them and fit within their preferred learning style. The following lists just some of the options that are particularly useful for geography:

- Colour code the paper used for the notes: knowledge, issues and case studies.
- Use highlighters, post-it notes and index cards to reduce information.
- For many geographical topics 'impacts' are an important consideration, subdivided into 'positive' and 'negative' or 'social/economic/environmental'. Colour coding of bullet point notes can be helpful when making lists of impacts.
- Geography lends itself to the use of many visuals in revision – diagrams of landform formations in physical geography, annotated photographs or sketch maps of case studies, charts, diagrams, tables and mind maps can be used to summarise notes and link ideas. The key is to break information down into manageable 'chunks'. These can then be inserted into diagrams of different shapes. Computer packages contain a wide range of SmartArt

graphics for text which adapt easily to geographical concepts – lists, process, cycle, hierarchy, relationship, pyramid and matrix. For examples, see Figures 5.1 and 5.2.

• New technologies can add much to revision strategies. Opportunities exist through online quizzes; YouTube and TeacherTube videos providing explanations; sites such as BBC Bitesize, and interactive whiteboard apps which allow teachers to record board work and audio explanations; online mind-mapping. Revision topics can also be presented in alternative formats such as PowerPoint and Chogger. In addition, laptops, tablets and smartphones allow students the flexibility to revise whenever they choose, which is an enormous bonus for motivation and can secure better outcomes.

Advantages and disadvantages: questions often asked about the impacts of an issue.

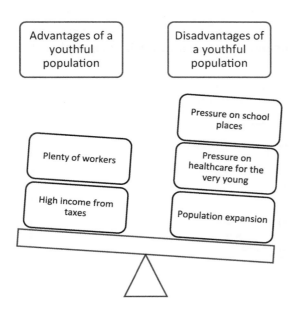

Opposing ideas: in geography both sides of an issue have to be taken into account.

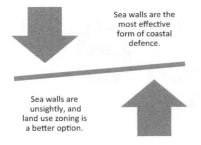

Figure 5.1 Useful graphics to aid revision in geography

Sequential steps: clear explanations in geography are achieved through explanations in the correct sequence of events.

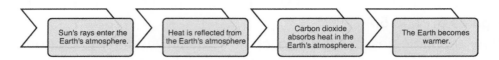

Building a case study: knowledge builds from descriptive facts to higher level analysis and evaluation of the case study.

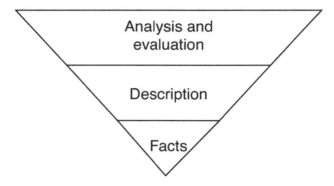

Cause and effect: this is a key format for explanations in geography.

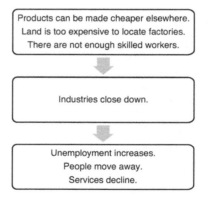

Figure 5.1 (Continued)

Figure 5.2 Revision word cloud for the Demographic Transition Model. Word clouds: these can be used to summarise a topic. Pupils can produce them as they revise, or the teacher can project one onto a whiteboard; questions then prompt revision, such as explaining the inclusion of certain words

Special consideration

For public and external examinations, schools need to apply for special arrangements to be put in place for pupils with SEND (www.jcq.org.uk/exams-office/access-arrangements-and-special-consideration). Schools will have to demonstrate that arrangements are required, often through an assessment by an educational psychologist. Support may include:

- extra time allocation for reading, writing and processing
- a separate room
- a reader (human or electronic)
- a scribe
- a prompter to keep focus
- use of an oral language modifier
- laptop use
- specialist software such as a screen reader
- supervised rest breaks
- use of coloured overlays or exam papers copied onto coloured paper
- specialist help such as explanation of key terms or help with colour interpretation for maps and data charts

Wherever possible, the pupil receiving support should have the opportunity to become familiar with the support strategy in key internal tests and examinations, e.g. a mock exam or end of year exam. It is also preferable if the role of reader and scribe is fulfilled by a member of staff with whom the recipient is comfortable and familiar.

Creating an effective assessment system

One of the key aims of the Assessment for Learning Strategy of 2008 stated that 'every school should have in place structured and systematic assessment systems for making regular, useful, manageable and accurate assessments of all pupils, and for tracking their progress.' Effective schools monitor and evaluate all pupils, including those with SEND, as part of a whole-school approach. In an inclusive school, it may well be rare for monitoring and assessment systems for pupils with SEND to be separate or different from those for other pupils. A positive feature of this approach is that assessment of pupils with SEND is very much the responsibility of all staff and not just that of the SENCO and support staff. However, for assessment to be effective for pupils with more complex needs, it may be necessary to build in the flexibility for an individualised approach where appropriate. Such an approach may involve using the P scales for pupils who are working below the standard of the national tests and assessments. In Key Stages 1, 2, 3 and 4, the P scales and performance descriptors in geography can be used as non-statutory guidelines describing the types and range of performance that pupils with SEND could demonstrate.

Geography Performance (P) scales:

P1 (i) Pupils encounter activities and experiences.
P1 (ii) Pupils show emerging awareness of activities and experiences.
P2 (i) Pupils begin to respond consistently to familiar people, events and objects.
P2 (ii) Pupils begin to be proactive in their interactions.
P3 (i) Pupils begin to communicate intentionally.
P3 (ii) Pupils use emerging conventional communication.
P4 Pupils extend the skills to help them explore the world.
P5 Pupils consolidate a sense of place and direction [for example, they can follow set routes around familiar places].
P6 Pupils understand the differences between the physical/natural and human/made features of places.
P7 Pupils communicate their preferences about the physical/natural and human/made features of places.
P8 Pupils recognise the physical/natural and human/made features of places [for example, identifying buildings and their uses].

www.gov.uk/government/uploads/system/uploads/attachment_data/file/329911/Performance_-_P_Scale_-_attainment_targets_for_pupils_with_special_educational_needs.pdf

For schools to develop an inclusive, whole-school approach to monitoring and assessment there must be a commitment to professional development and an input of time and resources. A useful starting point is to look at where the school is at and how progress can be made to establish an effective assessment system for all pupils.

The whole-school policy must then be adopted by all departments and individual teachers in order to become fully 'integrated' (Figure 5.3).

Motivation

Assessment of all kinds can be stressful – especially for pupils with SEND, some of whom make progress slowly and are painfully aware that others are leagues ahead. Keeping these learners motivated is essential in supporting them to reach their potential. Table 5.6 suggests how teachers can counteract any negative consequences of assessment, nurture pupils' self-esteem and help sustain their motivation.

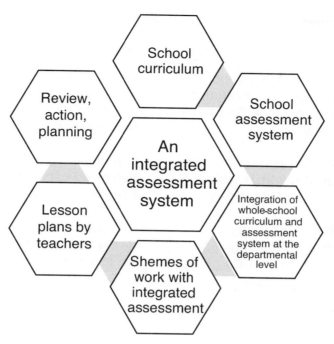

Figure 5.3 An integrated assessment system

To avoid the negative impact of tests on motivation for learning, teachers should:

Table 5.6 Implications for the work of teachers in the geography classroom

do more of this . . .	*and do less of this . . .*
Provide choice, when appropriate, and help students to take as much responsibility as they can for their learning.	Defining the curriculum in terms of what is in the tests/exams to the detriment of what is not tested.
Discuss with students, using appropriate communication strategies, the purpose of the learning and link new learning with previous learning/experience.	Giving frequent 'drill practice' for test taking.
Provide feedback that will help the learning process.	Emphasising competition for marks or grades among students, and using tests and assessments to tell students where they are in relation to others.
Encourage students to communicate (in an appropriate medium) how much they have learnt and what progress they feel they have made.	Teaching how to answer specific test questions.

(Continued)

Table 5.6 (Continued)

do more of this . . .	and do less of this . . .
Develop students' understanding of the goals of their work (success criteria), where they are and how to make further progress (next steps).	Allowing students to judge their work only in terms of scores or grades and become anxious about their performance.
Encourage students to value effort and a wide range of attainments by using different and appropriate strategies.	Giving feedback relating to students' perceived capabilities, implying a fixed view of each student's potential.
Encourage collaboration among students by using appropriate groups/ pairings; this will help to develop a positive view of each others' attainments.	Comparing students' grades, giving status on the basis of test achievement only.

6 Managing support

Support for pupils with SEND may take many forms; there are resources such as special equipment, adapted texts and activity sheets, and technology – there are references to these in Chapters 2 and 3. This chapter will concentrate on managing the human resources – the additional adults known as TAs or support/learning/inclusion/curriculum assistants as they are variously described.

For several years, the number of TAs and other support staff in schools has been rising significantly: there were over 250 000 FTE (full-time equivalent) TAs and 230 000 FTE support staff in state funded schools in 2014, with these two groups together outnumbering qualified teachers (DfE National Statistics, November 2014). The trend for employing an ever-increasing number of TAs, coupled with a tendency for teachers to rely on them for 'taking care' of pupils with SEND, has recently been stemmed, however, partly because of funding but also because of research findings. During the DISS study (Deployment and Impact of Support Staff: 2003–2008), for example, researchers found that TA involvement has not always been shown to improve outcomes for students; in fact, in some cases, there has been a negative effect (Blatchford et al., 2009). There is evidence that a misguided, albeit well-meaning TA can nurture dependency in a pupil, act as a barrier between him/her and the qualified teacher, and significantly limit learning and progress – especially in situations where a pupil interacts primarily with one or several TAs throughout much of the school day.

Subsequent projects (see Webster et al., 2016) have made a convincing argument for schools to reassess how they recruit, train and deploy these paraprofessionals. It is clear that with the right sort of knowledge, experience and attitude, TAs can and do play a major role in meeting students' individual needs and maximising their potential. Those employed to support individual pupils (usually those who have significant needs and therefore often an education, health and care (EHC) plan) will know the individual very well, and will be a useful source of information for subject teachers, about exactly what the pupil can and cannot do, the strategies that work well and any specialised resources that can be used.

TAs can also:

- encourage pupils to remain focused and on task
- monitor behaviour and reduce the incidence of disruption
- foster greater pupil independence (encourage pupils to 'have a go')
- increase pupils' confidence and self-esteem
- clarify and explain instructions, questions and tasks
- keep pupils on task
- oversee the setting up and care of equipment
- read or help pupils to read written material
- differentiate tasks and resources
- encourage and praise pupils
- help pupils to work towards their individual targets

The appropriate deployment of support underpins success or failure in many a classroom. Teachers of geography should be supported by their Head of Department, who in turn needs to be supported by senior management, in adopting flexible approaches and schemes of work that break down the barriers to learning. Supporting students is clearly not just about curriculum coverage. It is about developing strategies to access geographical content in a stimulating way; it is about developing enquiring minds and supporting students to ask geographical questions and investigate their own world thoughtfully and appropriately.

Achieving good practice

1 The basics

- Introduce your TA and promote them as a valued member of staff who deserves co-operation and respect. The students will see you as a team, and the TA will feel valued and confident.
- Establish and negotiate joint working practices: routines, expectations of behaviour, limits of flexibility. Avoid situations where teacher and TA are responding in different ways to pupils' questions and requests, e.g. leaving a lesson for going to the toilet, or for time out.
- Agree on confidentiality guidelines; avoid discussing pupils with SEND in the hearing of other pupils, for example.

2 Developing a workable relationship

- TAs need to know the lesson content. Provide copies of schemes of work so they know the aims and objectives of what you are doing.

- Establish a regular meeting time to go over your lesson plans and share any special requirements for individual pupils.
- Tell TAs exactly what you want them to do, with whom, when and for how long.
- Provide examples of answers, good pieces of work, etc., to inform TAs' expectations of the pupils they are supporting.
- Let your TA know if there is a problem, something that you feel needs to change. Be tactful but honest, and give them opportunities to respond. A TA should feel able to approach you about anything that is concerning him/her, or something they're not sure about.
- Review lessons together, and give the TA a role in assessing pupils' progress and achievement.
- Acknowledge the efforts of a TA and remember to show appreciation so that she/he feels valued.

If you have support for fieldwork, make sure the TA is briefed appropriately and understands the learning objectives

Securing and developing TA support in geography lessons

Schools organise TA support in different ways according to different criteria. You should familiarise yourself with how your school allocates its support and understand the 'bidding process'. There will be pupils with EHC plans who have a specific allocation of support during the school day – perhaps in certain lessons – but many others without EHC plans who would benefit from an extra adult in the classroom. Geography teachers who can demonstrate effective deployment of TAs and an understanding of how their support can raise achievement will be best placed to secure an allocation of TA time.

Some schools allocate TAs to particular subject departments or faculties, and this can be an advantage in several ways:

- Pupils learn to interact with several TAs rather than becoming dependent on just one person.
- The TA learns the subject material and becomes familiar with teaching methods (and is able to provide continuity when supply staff are brought in).
- Teacher/TA relationships develop well and underpin good practice; sometimes, the TA is able to supervise the class whilst the teacher works with pupils who need extra explanation and reinforcement.
- With experience, the TA is able to give valuable insights and tips on differentiation. He or she may develop a bank of resources – differentiated tasks, simplified texts, visual support material.

Where TAs are allocated to the Geography department, it will pay dividends to provide some subject training: you could consider a regular 'topic introduction' session, explaining terms and main teaching points, and providing copies of textbooks and handouts for preparation before lessons. Foster relationships whereby a TA always feels able to ask questions, without feeling embarrassed or foolish – even if this is in the presence of pupils. It is far better to admit not knowing something than to 'bluff it' and risk misleading pupils.

Whether or not a TA specialises in Geography, he or she should have detailed knowledge of individual pupils with SEND and understand their learning needs. The TA should be familiar with individual learning and behaviour targets and be able to remind the pupil of them as well as monitoring progress made with these. Close liaison with the SENCO, and sometimes with outside agencies, is also important. TAs can make valuable observations during lessons and contribute to individual plans, assessments and reviews when afforded appropriate training and opportunity. They should also become skilled in identifying pupils who seem to be having difficulties and who may have 'slipped through the net' of formal identification procedures. A TA may also provide effective pastoral support for pupils; having a good understanding of their personal and learning needs, they are able to encourage and motivate individuals and act as a sounding board when things don't always go well. The essential point here is that the TA maintains a professional stance and avoids sympathising with the pupil at the expense of the teacher.

Ryan is in Year 7. Chris, his TA, ensures he carries spare board pens with him in case the teacher runs out. She checks that he has his sloping board each day and goes with him once a term to see his Visual Impairment Advisory Teacher. In geography, Chris ensures Ryan has access to his CCTV enlarger, which she books out from the SEN department particularly for image or map work.

Katie is in Year 11 and also has a visual impairment. She needs all work enlarged from A4 to A3. Chris accompanies Katie on all fieldwork, both inside the school grounds and on field trips away from school, for health and safety reasons.

Deployment of TAs

Management of other adults is now an established part of a teacher's role. It can easily come to represent 'more work', however, and slip down to the bottom of one's priorities list. The planning sheet provided (Appendix 6.1 and online) may help – it's a quick way of communicating with a TA about specific roles in any lesson and offers an aide-memoire of the possibilities (highlight appropriate tasks and insert details in the right-hand column). The secret of high quality support in the classroom, however, lies in the relationship you develop with TAs: good communication, mutual respect and consideration, acknowledgement of their efforts and praise when things go well will enable you to establish working practices that result in truly effective teaching and learning.

Skilled TAs will know the pupils with SEND, know the subject matter and know how the teacher operates in the classroom (and out in the field). The examples in the following outline different types of effective practice.

At the start of a geography lesson

Initial learning

The teacher could support the assistant and student by	*The assistant could support the teacher and student by*
Indicating on the lesson plan which enquiry process is to be followed	Asking leading questions, drawing in reticent students
Providing names of pupils who need support with specific elements of the lesson	Helping the teacher manage tasks by issuing resources and observing and recording pupils' contributions
Checking on the progress of pupils with SEND throughout the lesson	Re-focusing a student's attention
Setting agreed time checks within the lessons	Giving feedback to the teacher about pupils who haven't finished, and any stumbling blocks

For example, when analysing photographs of Nairobi, the TA produced a laminated spider diagram with a spinner on it. The spinner was spun, and it stopped on one of ten key geographical enquiry questions to help the students analyse the image.

During a simple introductory activity for a group of Year 7 pupils studying volcanoes the TA read what the teacher had used the year before and simplified it for a small group of students with numeracy difficulties. The pupils had to decide which ten things they would take from their house to the emergency shelter once the eruption warning siren was heard. Each of the 25 options had a weight, and their rucksack could carry 20 kgs. The TA simplified the weights to multiples of 2, 5 and 10 and checked that the students would have the vocabulary knowledge to choose from all 25 options.

During the main teaching period

During the main learning

The teacher could support the assistant and student by	The assistant could support the teacher and student by
Providing the TA with answers to questions or opposing arguments or main points in discussion work (before the lesson)	Enacting the other viewpoint in a role play, playing devil's advocate in a debate
Ensuring the TA knows where all resources are stored	Acting as a scribe for a student, or for the teacher while students volunteer answers
Ensuring the TA knows the aims and objectives for the lesson, providing them with their own copies of handouts etc.	Re-directing reading, encouraging and supporting individuals. Pre-empting and addressing misconceptions. Helping students check their work
Ensuring the TAs know what has to be achieved by different individuals	Praising students as appropriate. Questioning students about what they have learnt – preparing for the plenary

For example, during a lesson on changing employment structure the TA recorded the jobs that the students' parents did. She recorded them in three different colours for primary, secondary and tertiary to help the class work out the differences. Later, during the write-up, she used the teacher's simplified flow diagram to help the students classify a list of other jobs and then volunteered her own family example to show how employment opportunities have changed over time in Stoke on Trent.

Towards the end and after the lesson

Closing thoughts/plenary

The teacher could support the assistant and student by	The assistant could support the teacher and student by
Setting up expectations that all students will contribute, as appropriate	Prompting the students to supply answers, praising and encouraging
Ensuring the student believes good work will be recognised with merits or credits or positive comments	Mentioning the good work a particular student has done
Ensuring homework is explained in good time – not set quickly in the last minute of the lesson	Noting down the homework for the student, checking they know what to do
Asking for the TA's views on progress	Informing the teacher of progress/misconceptions
Ensuring the TA feels confident and valued	Being involved in evaluating this lesson and planning and preparing resources for the next

For example, the TA suggested that a writing frame that gave more support to David might be needed to help him extend his writing. The class had been studying migration in Italy. At the moment he could only think of the effects on those already living in Turin when discussing Guiseppe's move from Aliano. He had not grasped the consequences for Guiseppe's own town. David had used the frame well to write about two effects on the people and two on the environment, but he had not made any other links.

The suggestions for TA deployment (Appendix 6.1 and an online resource) can be used by a teacher to indicate specific tasks for a lesson. It can also be used to structure a departmental session on the effective and purposeful use of support staff. Both teachers and TAs could annotate this list with geography-specific examples as to how they can modify their work to reduce the barriers to learning in geography.

Supporting pupils in a lesson about the water cycle

The lesson focus was to learn about and develop an understanding of the water cycle. Appendix 6.2 was used (this also appears as an online resource) – a set of sentences to be cut up into cards to sequence. The key words for this sequence of learning were identified:

Evaporation
Condensation
Cycle
Extension – Weathering
Erosion

The teacher communicated to the TA that, as a minimum, he expected all students to know and understand the key words. He expected that the majority would be able to sequence the events of the water cycle correctly and that some students might access some extension activities relating to erosion and weathering. A few students might start to make connections between weathering, erosion and the water cycle. It was agreed that the TA would work in the following ways to enable pupils with SEN to achieve the learning objective and record their work:

- provide help with looking up key words in the dictionary
- number the cards to enable self-checking by the pupils
- use paired reading to access information on the cards in the envelope
- use information from the cards and the picture as a base to illustrate the water cycle
- help pupils to become familiar with the key words through paired talk and a question-and-answer session
- focus attention on key points and set a limit on the number of sentences to be written, i.e. five sentences on the water cycle

Summary

Working effectively with colleagues to plan and implement an appropriate geography curriculum can have very positive benefits to learning for all pupils, but especially those with SEND. Relationships are key: respect and mutual understanding are often strongly evident in effective geography classrooms. Purposeful planning and liaison are also features of successful learning environments, and teachers are encouraged to think with clarity about the key learning objectives of the lesson and to share these with TAs.

Appendix 1.1 Legislation and guidance

The Children and Families Act: a different landscape

The Children and Families Act 2014 introduced radical changes to the requirements placed on both schools and teachers regarding the education and inclusion of pupils with special educational needs and disabilities (SEND). The first major revision of the SEN framework for 30 years, it introduced a new system to help children with SEND and shaped how education, health and social care professionals should work in partnership with children, young people and their families.

The reforms introduced a system to support children and young people from birth up to the age of 25, designed to ensure smooth transitions across all services as they move from school into further education, training and employment. The reforms give particular emphasis to preparing children and young people for adulthood from the earliest years. This means enabling children to be involved at as young an age as possible in all decisions relating to their learning, therapy, medical treatment and support from social care. The result of this preparation should be that when young people reach the age of 16, they are able to be full and active participants in all important decisions about their life.

> *There is now an important distinction made between a child and a young person. The Act gives significant new rights directly to young people when they are over compulsory school age but under the age of 25. Under the Act, a child becomes a young person after the last day of summer term during the academic year in which he or she turns 16. This is subject to a young person 'having capacity' to take a decision under the Mental Capacity Act 2005.*

Throughout this book the term 'pupils with SEND' is used. A pupil has SEN if he or she:

- has a significantly greater difficulty in learning than the majority of others of the same age; or
- has a disability which prevents or hinders him or her from making use of facilities of a kind generally provided for others of the same age in mainstream schools or mainstream Post-16 institutions.

(SEND Code of Practice 2015)

Section 19 principles

Central to Part 3 of the Children and Families Act 2014 is Section 19. This section emphasises the role to be played by parents/carers and young people themselves, in all decision-making about their SEND provision.

Part C of Section 19 issues a new challenge to schools in that there is a clear expectation that parents and pupils not only will be invited to participate but should be supported to do so. This will certainly involve the provision of relevant information to parents, but schools could also consider providing other forms of support: both practical support, such as helping with translation services, or even transport to attend important meetings, and emotional support, such as advocacy or pre-meetings to prepare parents and pupils to take a full part in all decisions. Many parents will need only a minimal level of additional support, but others – especially those often portrayed as 'hard to reach' – may require considerably more.

Key questions

- Do you know the wishes and feelings about education of your pupils with SEND and their parents? If not, how can you find out?
- What could you and others in your subject/departmental team do to integrate this information into your planning for and delivery of teaching and learning?
- What more could you do to reach out to parents who may be anxious about or unwilling to engage with school?

The SEND Code of Practice

As the quotation at the start of this appendix makes clear, the SEN provision is a provision that is additional to or different from the high quality, differentiated teaching to which all pupils are entitled. A school's first response to a pupil falling behind his/her peers should be to evaluate the quality of teaching and learning the pupil currently receives in all subjects. The pupil should be identified as having SEN only when the school is confident that all teaching is differentiated appropriately to meet that individual pupil's needs.

Once a pupil is identified as having SEN, schools are required to do whatever they can to remove any barriers to learning and to put in place effective provision, or 'SEN Support'. This support must enable pupils with SEN to achieve the best possible outcomes.

Most schools and academies welcome pupils with a range of vulnerabilities, including SEND, but may hesitate about including those with significant or complex needs. The reasons behind this reluctance are often a lack of expertise in an area of need, worries about behaviour, and, most commonly expressed, concerns about the impact of that pupil's needs on the education of others.

The SEND Code of Practice is very clear that where the parent of a pupil with an Education, Health, and Care plan (EHC plan) makes a request for a particular school, the local authority *must* comply with that preference and name the school in the plan unless:

- it would be unsuitable for the age, ability, aptitude or SEN of the child or young person, or
- the attendance of the child or young person there would be incompatible with the efficient education of others, or the efficient use of resources.

(SEND Code of Practice 2015, 9.79, p. 172)

Legally, schools cannot refuse to admit a pupil who does not have an EHC plan because they do not feel able to cater for his/her needs, or because the pupil does not have an EHC plan.

Outcomes

Outcomes are written from the perspective of the pupil and should identify what the provision is intended to achieve. For example, do you think the following is an outcome for a pupil in Year 7 with literacy difficulties?

For the next ten weeks Jake will work on an online literacy program for 20 minutes three times each week.

It may be specific and measureable; it is achievable and realistic; and it is time targeted, so it is 'SMART', but it isn't an 'outcome'. What is described here is a provision, i.e. the intervention that the school will use to help Jake to make accelerated progress.

Outcomes are intended to look forward to the end of the next stage or phase of education, usually two or three years hence. Teachers will, of course, set short-term targets covering between 6 and 12 weeks, and education, health and care plans will also include interim objectives to be discussed at annual reviews. So, what would be an outcome for Jake?

> By the end of Year 9, Jake will be able to read and understand the text-books for his chosen GCSE courses.

The online literacy course would then form a part of the package of provision to enable Jake to achieve this outcome.

The graduated approach

The SEND Code of Practice describes SEN Support as a cyclical process of assess, plan, do and review that is known as the 'graduated approach'. This cycle is already commonly used in schools, and for pupils with SEN it is intended to be much more than a token, in-house process. Rather, it should be a powerful mechanism for reflection and evaluation of the impact of SEN provision. Through the four-part cycle, decisions and actions are revisited, refined and revised. This then leads to a deeper understanding of an individual pupil's needs whilst also offering an insight into the effectiveness of the school's overall provision for pupils with SEN. The graduated approach offers the school, the pupil and his/her parents a growing understanding of needs and of what provision the pupil requires to enable him/her to make good progress and secure good outcomes. Through successive cycles, the graduated approach draws on increasingly specialist expertise, assessments and approaches, and more frequent reviews. This structured process gives teachers the information they need to match specific, evidence-based interventions to pupils' individual needs.

Evidence-based interventions

In recent years, a number of universities and other research organisations have produced evidence about the efficacy of a range of different interventions for vulnerable pupils and pupils with SEN. Most notable among this research is that sponsored by the *Education Endowment Fund* that offers schools valid data on the impact of interventions and the optimal conditions for their use. Other important sources of information about evidence-based interventions for specific areas of need are the *Communication Trust – What Works?* website and *Interventions for Literacy* from the *SpLD/Dyslexia Trust*. Both sites offer transparent and clear information for professionals and parents to support joint decisions about provisions.

The Equality Act 2010

Sitting alongside the Children and Families Act 2014, the requirements of the Equality Act 2010 remain firmly in place. This is especially important because many children and young people who have SEN may also have a disability under the Equality Act. The definition of disability in the Equality Act is that the child or young person has 'a physical or mental impairment which has a long-term and substantial adverse effect on a person's ability to carry out normal day-to-day activities'.

'Long-term' is defined as lasting or being likely to last for 'a year or more', and 'substantial' is defined as 'more than minor or trivial'. The definition includes sensory impairments such as those affecting sight or hearing, and, just as crucially for schools, children with long-term health conditions such as asthma, diabetes, epilepsy and cancer.

As the SEND Code of Practice (DfE, 2015, p. 16) states, the definition for disability provides a relatively low threshold and includes many more children than schools may realise. Children and young people with some conditions do not necessarily have SEN, but there is often a significant overlap between disabled children and young people, and those with SEN. Where a disabled child or young person requires special educational provision, they also will be covered by the SEN duties.

The Equality Act applies to all schools, including academies and free schools, university technical colleges and studio schools, and also further education colleges and sixth form colleges – even where the school or college has no disabled pupils currently on roll. This is because the duties under the Equality Act are anticipatory in that they cover not only current pupils but also prospective ones. The expectation is that all schools will be reviewing accessibility continually and making reasonable adjustments in order to improve access for disabled pupils. When thinking about disabled access, the first thing that school leaders usually consider is physical access, such as wheelchair access, lifts and ramps. But physical access is only part of the requirements of the Equality Act and often is the simplest to improve. Your school's accessibility plan for disabled pupils must address all of three elements of planned improvements in access:

1 physical improvements to increase access to education and associated services
2 improvements in access to the curriculum
3 improvements in the provision of information for disabled pupils in a range of formats

Improvements in access to the curriculum are often a harder nut to crack as they involve all departments and all teachers looking closely at their teaching and learning strategies and evaluating how effectively these meet the needs of disabled pupils. Often, relatively minor amendments to the curriculum or teaching approaches can lead to major improvements in access for disabled pupils, and these often have a positive impact on the education of all pupils. For example, one school installed a Soundfield amplification system in a number of classrooms because a pupil with a hearing loss had joined the school. The following year, the cohort of Year 7 pupils had particularly poor speaking and listening skills, and it was noticed that they were more engaged in learning when they were taught in the rooms with the Soundfield system. This led to improvements in progress for the whole cohort and significantly reduced the level of disruption and off-task behaviours in those classes.

Schools also have wider duties under the Equality Act to prevent discrimination, to promote equality of opportunity and to foster good relations. These duties should inform all aspects of school improvement planning from curriculum design through to anti-bullying policies and practice.

Significantly, a pupil's underachievement or behaviour difficulties might relate to an underlying physical or mental impairment which could be covered by the Equality Act. Each pupil is different and will respond to situations in his/her unique way so a disability should be considered in the context of the child as an individual. The 'social model' of disability sees the environment as the primary disabling factor, as opposed to the 'medical model' that focuses on the individual child's needs and difficulties. School activities and environments should be considered in light of possible barriers to learning or participation.

Appendix 1.2 Departmental policy

Whether the practice in your school is to have separate SEND policies for each department or to embed the information on SEND in your whole-school Inclusion or Teaching and Learning policies, the processes and information detailed here will still be relevant.

Good practice for pupils with SEND and disabilities is good practice for all pupils, especially those who are 'vulnerable' to underachievement. Vulnerable groups may include looked-after children (LAC), pupils for whom English is an additional language (EAL), pupils from minority ethnic groups, young carers, and pupils known to be eligible for free school meals/Pupil Premium funding. Be especially aware of those pupils with SEND who face one or more additional vulnerabilities and for whom effective support might need to go beyond help in the classroom.

It is crucial that your departmental or faculty policy describes a strategy for meeting pupils' SEN within your particular curricular area. The policy should set the scene for any visitor, from supply staff to inspectors, and make a valuable contribution to the departmental handbook. The process of developing a departmental SEND policy offers the opportunity to clarify and evaluate current thinking and practice within the geography team and to establish a consistent approach.

The SEND policy for your department is a significant document in terms of the leadership and management of your subject. The preparation and review of the policy should be led by a senior manager within the team because that person needs to have sufficient status to be able to influence subsequent practice and training across the department.

What should a departmental policy contain?

The starting points for your departmental SEND policy will be the whole-school SEND policy and the SEND Information Report that, under the Children and

Families Act 2014, all schools are required to publish. Each subject department's own policy should then 'flesh out' the detail in a way that describes how things will work in practice. Writing the policy needs to be much more than a paper exercise completed merely to satisfy the senior management team and OFSTED inspectors. Rather, it is an opportunity for your staff to come together as a team to create a framework for teaching geography in a way that makes your subject accessible, not only to pupils with SEND, but to all pupils in the school. It is also an ideal opportunity to discuss the impact of grouping on academic and social outcomes for pupils. (Bear in mind that the Code of Practice includes a specific duty that '*schools must ensure that pupils with SEND engage in the activities of the school alongside pupils who do not have SEND*' (DfE, 2015, 6.2, p. 92, emphasis added).

We need to be careful in geography that when grouping pupils, we are not bound solely by measures in reading and writing, but also take into account reasoning and oral language abilities. It is vital that social issues are also taken into account if pupils are to be able to learn effectively. Having a complement of pupils with good oral ability will lift the attitude and attainment of everybody within a group.

Who should be involved in developing our SEND policy?

The job of developing and reviewing your policy will be easier if tackled as a joint endeavour. Involve people who will be able to offer support and guidance such as:

- the school SEND governor
- the SENCO or other school leader with responsibility for SEND
- your support staff, including TAs and technicians
- the school data manager, who will be able to offer information about the attainment and progress of different groups
- outside experts from your local authority, academy chain or other schools
- parents of pupils with SEND
- pupils themselves – both with and without SEND

Bringing together a range of views and information will enable you to develop a policy that is compliant with the letter *and* principle of the legislation; that is relevant to the context of your school; and that is useful in guiding practice and improving outcomes for all pupils.

The role of parents in developing your departmental SEND policy

As outlined in Appendix 1.1, Section 19 of the Children and Families Act 2014 raises the bar of expectations about how parents should be involved in and influence the work of schools. Not only is it best practice to involve parents of pupils with SEND in the development of policy, but it will also help in 'getting it right' for both pupils and staff. There are a number of ways, both formal and informal, to find out the views of parents to inform policy writing, including:

- focus group
- coffee morning/drop-in
- questionnaire/online survey
- phone survey of a sample of parents

Parents will often respond more readily if the request for feedback or invitation to attend a meeting comes from their son or daughter.

Where to start when writing a policy

An audit can act as a starting point for reviewing current policy on SEND or writing a new policy. This will involve gathering information and reviewing current practice with regard to pupils with SEND, and is best completed by the whole department, preferably with some input from the SENCO or another member of staff with responsibility for SEND within the school. An audit carried out by the whole department provides a valuable opportunity for professional development so long as it is seen as an exercise in sharing good practice and encourages joint planning. It may also facilitate your department's contribution to the school provision map. But before embarking on an audit, it is worth investing some time in a departmental meeting, or ideally a training day, to raise awareness of the legislation around SEND and to establish a shared philosophy across your department.

The following headings may be useful when you are establishing your departmental policy:

General statement of compliance

- What is the overarching aim of the policy? What outcomes do you want to achieve for pupils with SEND?
- How are you complying with legislation and guidance?
- What does the school SEND Information Report say about teaching and learning and provision for pupils with SEND?

Example

All members of the department will ensure that the needs of all pupils with SEND are met, according to the aims of the school and its SEND policy . . .

Definition of SEND

- What does SEND mean?
- What are the areas of need and the categories used in the Code of Practice?
- Are there any special implications for our subject area?

Provision for staff within the department

- Who has responsibility for SEND within the department?
- What are the responsibilities of this role?

e.g.

- liaison between the department and the SENCO
- monitoring the progress of and outcomes for pupils with SEND, e.g. identifying attainment gaps between pupils with SEND and their peers
- attending any liaison meetings and providing feedback to colleagues
- attending and contributing to training
- maintaining departmental SEND information and records
- representing the needs of pupils with SEND at the departmental level
- liaising with parents of pupils with SEND
- gathering feedback from pupils with SEND on the impact of teaching and support strategies on their learning and well-being

(This post can be seen as a valuable development opportunity for staff, and the name of this person should be included in the policy. However, where responsibility for SEND is given to a relatively junior member of the team, there must be support and supervision from the head of the department to ensure that the needs of pupils with SEND have sufficient prominence in both policy and practice.)

- What information about pupils' SEND is held, where is it stored, and how is it shared?
- How can staff access additional resources, information and training?
- How will staff ensure effective planning and communication between teachers and teaching assistants?
- What assessments are available for teachers in your department to support accurate identification of SEND?

Example

The member of staff with responsibility for overseeing the provision of SEND within the department will attend liaison meetings and subsequently give feedback to the other members of the department. She/he will maintain the department's SEND file, attend and/or organise appropriate training and disseminate this to all departmental staff. All information will be treated with confidentiality.

Provision for pupils with SEND

How are pupils' special educational needs identified?
e.g.

- observation in lessons
- assessment of class work/homework
- end of module tests/progress checks
- annual examinations/SATs/GCSE
- reports

- How is progress measured for pupils with SEND?
- How do members of the department contribute to individual learning plans, meetings with parents and reviews?
- What criteria are used for organising teaching groups?
- How/when can pupils move between groups?
- What adjustments are made for pupils with SEND in lessons and homework?
- How do we use information about pupils' abilities in reading, writing, speaking and listening when planning lessons and homework?
- What alternative courses are available for pupils with SEND?
- What special arrangements are made for internal and external examinations?
- What guidance is available for working effectively with support staff?

Here is a good place also to put a statement about the school behaviour policy and any rewards and sanctions, and how the department will make any necessary adjustments to meet the needs of pupils with SEND.

Example

The staff in the (subject) department will aim to support pupils with SEND to achieve the best possible outcomes. They will do this by supporting pupils to achieve their individual targets as specified in their individual learning plans, and will provide feedback for progress reviews. Pupils with SEND will be included in the departmental monitoring system used for all pupils.

Resources and learning materials

- Is any specialist equipment used in the department?
- How are differentiated resources developed? What criteria do we use (e.g. literacy levels)?
- Where are resources stored, and are they accessible for both staff and pupils?

> ## Example
>
> *The department will provide suitably differentiated materials and, where appropriate, specialist resources to meet the needs of pupils with SEND. Alternative courses and examinations will be made available where appropriate for individual pupils. Support staff will be provided with curriculum information in advance of lessons and will be involved in lesson planning. A list of resources is available in the department handbook.*

Staff qualifications and continuing professional development (CPD)

- What qualifications and experience do the members of the department have?
- What training has already taken place, and when? What impact did that training have on teaching and learning, and progress for pupils with SEND?
- How is training planned? What criteria are used to identify training needs?
- What account of SEND is taken when new training opportunities are proposed?
- Is a record kept of training completed and ongoing training needs?

> ## Example
>
> *A record of training undertaken, specialist skills and training required will be kept in the department handbook. Requests for training will be considered in line with the department and school improvement plan.*

Monitoring and reviewing the policy

- How will the policy be monitored?
- Who will lead the monitoring?
- When will the policy be reviewed?

> ## Example
>
> *The departmental SEND policy will be monitored by the Head of Department on a planned annual basis, with advice being sought from the SENCO as part of the three-year review process.*

Conclusion

Creating a departmental SEND policy should be a developmental activity that will improve teaching and learning for all pupils, but especially for those who are vulnerable to underachievement. The policy should be a working document that will evolve and change over time; it is there to challenge current practice and to encourage improvement for both pupils and staff. If departmental staff work together to create the policy, they will have ownership of it; it will have true meaning and be effective in clarifying good practice.

An example of a departmental policy for you to amend is available for download on the dedicated website.

Appendix 1.3 Creating a sensory experience

Geography is about increasing awareness of the wider world in all its diversity and beauty. Students with learning difficulties can develop a better understanding of different places and peoples when a sensory approach is employed to enhance their learning.

The sensory area/room can be set up to represent a place

For example, it could be:

- a tropical rainforest
- the seaside
- a Caribbean banana plantation
- an African market
- the middle of a big city

What you will need

- Images (from the internet) projected on the board.
- The room heated to the approximate temperature (and humidity) of the place (e.g. kettles boiled in the room to create tropical rainforest heat and humidity).
- The floor dressed with appropriate material (e.g. sand and pebbles for a beach).
- Some sounds of the place and/or music: downloaded sound effects from the internet (e.g. the sounds of a street in a big city).
- Smells recreated by cooking food or spraying aromas (e.g. the smell of fruit and vegetables of an African market). These can be tasted if appropriate too.
- Artefacts that can be handled, if appropriate (e.g. rocks, seaweed, driftwood from the beach).

The key is to think how the environment you are studying can be recreated.

Appendix 2.1 Tourism word mat

Places
Peak District
Disneyland
Paris
France
Greece
Kenya
Mediterranean
Majorca
Menorca
Spain
Turkey
USA

Adjectives
attractive
brash
coastal
cultural
historic
lively
mountainous
natural
quaint
quiet
scenic
spectacular
sunny
tropical
vibrant
wild

Problems
seasonal jobs
hotels owned by foreign
  owners
traffic congestion
increase in crime
woodlands cleared for
  development
noise
water shortages
increase in prices
visual pollution

Key words
activity holiday
advertising
airport
attractions
beach
biodiversity
brochure
climate
conflict
conservation
destination
ecotourism
honeypot
hotel
images
national park
package holiday
sightseeing
skiing
sustainability
tourism

Benefits
new jobs
investment
improved roads
foreign income
sustainable tourism
better services
new roads and airports

Appendix 3.1 Case studies

Pupils with different types of SEND

Mary: Asperger's syndrome

Callum: Attention deficit disorder/attention deficit hyperactivity disorder (ADD/ADHD)

John: Autism/autistic spectrum disorder (ASD)

Gus: Cerebral palsy (CP)

Susan: Down's syndrome

Saul: Fragile X syndrome

Yasmin: Moderate learning difficulty/disability

Nicky: Severe learning difficulty/disability

Alice: Profound and multiple learning difficulties

Louise: Visual impairment

Paul: Hearing impairment

Lizzie: Multi-sensory impairment

Sally: Social, emotional and mental health needs

Sam: Dyslexia

Shajid: Dyscalculia

James: Dyspraxia

Chantelle: Speech, language and communication needs

Michael: Tourette's syndrome

Mary

My name is Mary. I am 15 years old and have a diagnosis of Asperger's syndrome. When I was small I could never understand other children's behaviour. I would watch them playing the same favourite game many times but could never understand the rules and therefore could never join in. As I grew up I became even more confused about social interactions; although I have tried to learn the rules they are always changing and often appear illogical. Unfortunately, I am unable to interpret non-verbal communication and only hear words; therefore, I have great difficulty following a conversation, listening, being tactful, taking hints, making small talk and knowing what is acceptable to say. Consequently, I am always upsetting people, and have few friends.

Other people also think I am rude or don't listen, but once I am focused on something I cannot just think about something else.

Callum

My name is Callum, and I am 13 years old and have a diagnosis of ADHD. I take medication to help me to control my feelings and to keep me calm. I try to make friends, but I find it difficult to do what they say, keep to the rules or wait my turn; then they don't want to know me and tell me to go away. Then I hit or kick; my feelings frighten me sometimes.

Sometimes I get very angry with my parents and teachers because they are always telling me what to do or think I am not listening because I am 'jigging' about, but for me this is how I listen the best. They don't understand how difficult it is to sit still and then listen to what they say. I suppose they are right most of the time; it's just that I get so excited about things, and I can't wait to try them or share what I know or have found out. My Mum says I do or say things before I put my mind in gear and that is why I get into lots of trouble. Another thing that drives my family and teachers mad is that I am always losing or forgetting things; what they don't realise is that this makes me as angry as they are.

I know that I am good at lots of things at school and have a very good memory, but it is not always easy for me to listen carefully for a long time. Sometimes my thoughts jump from one thing to another, and suddenly everything has moved on in a lesson and I am lost, have missed what I have been asked to do or answer the wrong questions. Some teachers think I do this on purpose; I don't, honestly.

John

I have ASD, which means I have special skills in a few areas. I know I have to do things in certain ways, and then I feel calm. I don't like school that much, but I do like geography. This is because I can look at maps and think about the motorways. Sometimes kids in my class shout at me because I like to make my special noise even though I do not always know when I am making it. I need things written down for me to understand what I have to do. It takes me longer to work out what people say to me, and sometimes teachers get angry because they think I am not listening or am refusing to do what I am told. I love clocks and how they work; sometimes I can talk about clocks in class because we talk about time zones. I like to go to my nan's. We go along three different motorways, and I calculate how many lamp-posts there are on each and work out the mileage average between them. When I get back, I look at the maps on the walls at school, and it makes me happy. I don't like people, especially girls, asking me about their clothes because I don't want to look at them and I know I can't get the answer right and it makes me sweat. I like to work on my own. I like working on the computer because it does not get angry.

Gus

Gus is a bright, enthusiastic student who has an opinion on everything and gives 100% to all he does at school. Gus has cerebral palsy, which makes him forgetful, a little disorganised at times and tired at the end of the day. Gus loses concentration as the day progresses and tires more quickly as the term progresses. Gus walks with an unsteady gait and struggles to control pens, pencils and rulers etc. as his fine motor skills are poor. Gus has to take his time to speak, and some students in his class can get irritated by this. Gus uses a laptop to type the majority of his school work and leaves lessons a few minutes early to get to his next lesson before the rest of the school population fills the corridors.

Gus receives support from all five of the school's teaching assistants during the week. He has an education, health and care plan which sets out strategies to ensure safety in school, to further develop his fine motor skills, to improve his spoken language and to make progress in all basic skills to have full access to the National Curriculum. These aims are to be achieved with any appropriate facilities or equipment, and any modifications to the curriculum. The school liaises with speech therapists and occupational therapists as necessary. An individual learning plan has been prepared by the school which details short-term targets for Gus and useful classroom strategies.

Background – rainforests

In the first lesson of the unit the class discussed and fed back what they already knew and understood about the rainforest. Gus, as always, had some prior knowledge and told the class about sloths and how they move slowly through the trees.

Strategies for Gus's inclusion

- Preparing the room before the class arrives so Gus does not have to move furniture
- Ensuring that he can sit next to a plug for his laptop
- Providing opportunities for Gus to share his ideas verbally
- Allowing Gus's teaching assistant to act as a scribe for the brainstorming, but giving Gus the responsibility of producing the final piece of work
- Setting the homework early in the lesson when Gus is always present

- Devising plenaries that check understanding without relying on a lot of student explanation
- Allowing Gus to leave early without missing the plenary

Susan

My name is Susan, I am nearly 16, and I have Down's syndrome. I don't 'suffer' from Down's syndrome, as many people seem to think; I was born that way, and I feel great about myself. I know that I have an extra chromosome and that I have some learning difficulties, but so do lots of my friends and they don't have Down's syndrome. I also have friends with Down's syndrome, and they are all different too.

I attend my local secondary school. I am good at reading but not so good at maths, and I use a computer to help me with my writing. Sometimes people who don't know me well find it difficult to understand what I am saying, but my family, friends and teachers don't have any problems at all.

Saul

Fifteen-year-old Saul has Fragile X syndrome and attends an integration unit on a mainstream school site. He has a great sense of humour, enjoys taking responsibilities and is a great favourite with adults. Saul appears to have no friends of his own age and has little contact or interaction with his peers. He is active and impulsive and often rushes into situations without thinking or being aware of what the consequences may be. Saul has a very good vocabulary and usually speaks in short phrases and sentences. He has a number of very amusing learned phrases that he uses in the correct context. He finds it difficult to answer direct questions and often gives answers that appear unusual or form part of his learned repertoire. Although Saul demonstrates that he enjoys verbal praise he appears to find it hard to accept and often responds by giggling or becoming extremely embarrassed. Saul becomes extremely anxious within social and group situations and will begin to chatter to himself using well-known phrases or instructions that he has been given and to repeat specific hand, head and body movements. These activities appear to bring the levels of anxiety down and have a calming effect. Lots of other situations throughout the school day raise his anxiety levels, including demanding class work, busy and noisy environments and changes in routines. He does not display any challenging behaviour within the school setting. Saul finds a structured environment, a visual and word timetable, clear instructions and adult modelling supportive to his learning.

In English and mathematics he is working at early National Curriculum levels and in other subjects P8 of P scales.

Yasmin

Yasmin has moderate learning difficulties. She is in Year 7. Her attention-seeking and avoidance behaviour can prove disruptive in geography lessons as she often shouts out or continually talks to her peers. Her short-term memory skills are good, and she can recall what she has been taught, but she has difficulty in sequencing information. Lots of practice and repetition help her to consolidate new concepts in geography. This strategy is also helpful to her when learning how to use appropriate specialist vocabulary. She has a reading age of 9. In mathematics she is working confidently with numbers up to 50, but she has difficulty generalising from familiar to new or practical situations. She likes to write but needs considerable support. She enjoys using the computer and works confidently with the program Clicker 6, which supports her reading, writing and recording. Yasmin states that her favourite lessons are 'when I can go outside and do things and work with my friends'. She enjoys geography field studies but requires lots of structure to organise materials and to help her to understand what is required of her and remain on task.

Nicky

Nicky has severe learning difficulties. He is 11 years old, and his education takes place within both a mainstream and a special school. He spends half his time in each setting. His mainstream class has the full-time support of a teaching assistant.

Nicky uses a mixture of single words, phrases and simple sentences, accompanied by signing, to communicate. Often sensory experiences are provided for Nicky. These link to the learning objectives of the curriculum planned for the rest of the class. He enjoys joining in with small group activities. Differentiated resources, content, and learning outcomes enhance his access to the curriculum. He has a sight vocabulary of 50+ words and uses Widget symbols (www.widget.com) to aid both his reading and writing. Nicky has a basic number concept to 10 and can complete simple addition and subtraction, and he applies these concepts in familiar practical and everyday situations.

Nicky can locate without support the main areas within the school and has been taught to use a simple pictorial and symbol map to find less familiar places that he does not access on a daily basis. He occasionally becomes confused with left and right although he uses his watch, always on his left wrist, to help him remember. He also has difficulties in identifying the position and location of objects in relation to himself, e.g. behind, by the side, near, far etc. He recognises basic differences and similarities in familiar but contrasting environments. In simple terms, when supported through prior learning and careful questioning, he can identify an environmental issue and suggest ways to improve the situation, e.g. litter in school grounds.

Alice

Alice's education, health and care plan identifies that she has 'difficulties in general learning skills, visual, communication, interaction, mobility and independence skills'. It is also noted that Alice has a visual impairment and intermittent conductive hearing difficulties. Assessment of Alice's level of achievement shows that she is presently working at a P2 (ii)/P3 (i) level in the core subjects.

Alice can present challenging behaviour, such as pulling hair, grabbing and pinching people who are in close proximity. This is often the result of unexpected changes or confusion. This behaviour has proved a useful strategy by which Alice can control invasions into her personal space. Alice has recently become an independent walker, and her confidence in her ability to move about within familiar areas is increasing.

Alice attends a special unit within a mainstream setting. She is a student within a class of nine, where the class teacher is supported by two teaching assistants. Alice has five hours of additional individual support from a key worker each week. The advisory teacher for the sensory impaired visits each half term, to support the work with Alice.

Alice's developing understanding of her school day is helped by a holistic approach to communication, incorporating the use of song, sign, objects of reference and familiar daily classroom routines.

All staff use a personal signal to alert Alice to who they are. They approach Alice in an unhurried and standardised way, in order to lessen her fear of the unexpected and to develop positive anticipation of human interaction. All staff members have an awareness of Alice's mobility needs. Signals are used to support Alice's understanding of changes in activity or location.

Alice's sensory impairment means that she may receive an incomplete or skewed view of the world, and many aspects of that environment may be unpredictable. In order to provide security for Alice and to support her understanding of place and function, the environment is structured, e.g. a personal work table in a specific and unchanging area of the classroom, and a relaxing area with access to favourite music etc. Additional communication approaches; physical, sensory and verbal cues; and a consistent approach by all staff provide Alice with the growing confidence to explore her immediate and wider environment.

Louise

Louise is a Year 7 student with a visual impairment. She has damaged optic nerves and reduced vision. Her sight is particularly poor at a distance, and bright light causes her discomfort and further reduces her vision. Louise is a very bright student who always seems to deal with her impairment in a matter of fact way, though at high school she has become more conscious of her impairment, and this has affected her self-confidence. Louise has also experienced some problems within her friendship group, due in part to her own insecurities.

Louise visited the school many times before her transition to help her find her way around the school site. She regularly sees the Visually Impaired Service, who monitor her inclusion and advise the school of any equipment which Louise could make use of. Louise currently uses a CCTV magnifier, Lunar ICT Enlarger software and a hand-held magnifier to assist her in class.

Louise's normal seat is at the front in all geography lessons, with the light coming into the room behind her. She was escorted from one lesson to another at the start of term but now knows her way around. Louise has support in practical subjects from teaching assistants for safety reasons alone; she has no support in geography.

Louise's class have been learning about the different ways that rocks are formed and the uses people have for different rocks in her recent geography lessons. Louise has brought her completed homework to the lesson; she has annotated an enlarged diagram of the rock cycle which checked her understanding of the formation of metamorphic rocks.

The objectives for this lesson were as follows below.

Students will:

- reinforce their knowledge of the three types of rocks – metamorphic, sedimentary and igneous
- learn how to use a classification system by sorting relevant from irrelevant information
- be able to describe what different rocks look like

Louise will:

- complete the same activities as the other students without a visual impairment
- share her views with others to boost her confidence
- realise that magnifying technology aids us all, not just those with an impairment. Use other senses besides sight to classify the rocks.

Additional support

- The teacher gave all instructions from the board, well away from the window, which distorts Louise's vision.
- All paper resources were enlarged twice the size – A5 to A4, A4 to A3.

- The teacher incorporated the CCTV enlarger into the lesson by using it to model how to examine the rock samples. Louise and her partner then both used her CCTV enlarger to view the rocks.
- The questions in the envelopes were printed on pale orange card to reduce the glare for everyone.
- The questions asked all students to touch the rocks, describe how they felt etc., which played to Louise's strengths.

Paul

Paul is 12 years old and attends his local high school. He has a sensorineural hearing impairment and wears hearing aids. He uses speech, but this is often stilted and unclear, especially when he is angry or excited. He uses BSL to support communication. He is a happy and highly motivated young man with a 'wicked' sense of humour. Although there have been difficulties in the past he has worked hard at developing his social and interactional skills and now has a close group of friends.

He sits at the front and to the left in class with the window behind him. This position enables him to make best use of his hearing on the right, supports his ability to lip read and access written information, and provides him with a clear view of his teacher.

Lizzie

Lizzie has a dual sensory impairment and a physical disability. She wears hearing aids and uses a walking frame to support her mobility. She has some useful peripheral vision that she uses effectively with touch to explore her environment. Lizzie is non-verbal but uses a range of strategies including vocalisation and gross motor movements to communicate her emotions, likes and dislikes, choices and responses to her environment. The staff use personal indicators, routines, objects of reference and enlarged black and white bold outlined key symbols to support meaning, communication and comprehension. The classroom is structured into clearly defined areas to facilitate the use of her residual vision and provide her with the security and confidence to explore and gain meaning from her environment.

Lizzie demonstrates that she is aware of patterns that form part of her daily routines. Within the classroom and wider internal and external school environment she can locate differentiated areas and associate familiar sensory stimuli, objects, activities, events and people to these areas. She enjoys accessing environments beyond the school and in particular enjoys the sensation of soft winds and the smell associated with wooded areas, and responds positively when revisiting these experiences through simulated sensory experiences.

Lizzie occasionally presents challenging and self-injurious behaviour when she is confused or frustrated. She is working at P 3(ii) in all areas of the National Curriculum. Her teacher has an additional qualification in the teaching of students with dual sensory impairment.

Sally

Sally is 13 years old and attends the local high school. She has experienced a very disruptive family life and has been in and out of care since the age of 4. At this time she lives with foster parents. Her attendance at school is poor. When she does attend she is often late and on entering a class will knock objects off the desks of other students and shout across the room to her friends. She refuses to listen in class or complete tasks set, and when requested to work she will either ignore the teacher or respond with abusive or threatening language. This behaviour is exhibited with all teachers at some time.

Individual behaviour management plans (IBMP) have been revised and adapted where appropriate in co-operation with the educational psychologists and foster parents. Teachers have worked closely together to share information and provide mutual support. The frequency of the outbursts has decreased over the past three months.

Sam's perspective on dyslexia

I would love to read age-appropriate stories about places and discover further information about my particular interests in geography, rocks! I am pleased that I have found part of the answer to this problem through the use of listening books. I find it very difficult to read and understand a written passage, but if someone else reads it to me and I follow the words I am able to answer the question much more quickly and waste less time. I hate trying to use punctuation, often get confused with upper and lower case letters and reverse some letters. I am really bad at spelling, and I usually learn how to spell a word by looking for patterns in the way the letters are arranged. This takes a long time, and I always forget how to spell the word when I need to use it in written work. This is extremely frustrating as I am able to remember other facts. I use a computer and spell check, which helps, but sometimes the word I have written is so different from the correct spelling that the wrong word is found.

My written work is often messy because I often have to cross things out when I get them wrong or lose my place. It annoys me when teachers judge me on my written work and reading and not on what I know. I get very tired by the end of the day and feel useless, and I am not.

Shajid

My name is Shajid, and I am a 15-year-old boy with dyscalculia. Many people have not heard about this, but it means that I have a mathematical learning disability, and only when you have this problem do you realise that numbers are everywhere. With anything involving numbers in geography, even simple adding and subtracting, I become very confused and make a great many mistakes. For example, I have a great problem with '0's', and if I am working with a number such as 301 I can see it in my mind, but if I need to write it down it becomes all mixed up, e.g. 310 or 130. If I am given geographical data verbally I am unable to picture the numbers in my head. If it is written down at least I have a chance, but sometimes even though I know how to solve a problem, or have a diagram to help me through the small steps of a task, I may miscopy the numbers, or place the answer next to the wrong number. I always check and double-check my answers, but I am unable to see that I have used the wrong sequence or term. Numbers are not the only things that I find difficult; reading the time on ordinary clocks, drawing shapes, following directions and identifying left and right remain a mystery. However, I have found some things that are helpful. These include using a digital watch and always wearing it on my left hand; this helps me with directions. I find plans and templates that help me know what to do next are useful. Teachers who support me in my use of a calculator and computer also help to improve my learning.

I am lucky because I do not also have problems with words like some of my friends with dyscalculia. I am also very good at practical things, and as my teachers say, I work hard, do my very best and have a great sense of humour. Working with my parents, teachers and friends I enjoy the challenge of finding different ways of beating my dyscalculia.

A typical day for James (dyspraxia)

It started with the ringing of the alarm, clothes in a muddle, shoes that would not stay tied. Late again and having forgotten a book. Trying hard to remember the number of things Mrs Smith had just asked us to do, leaned back to think, swung back too far and landed on the floor, everybody looking and laughing except Mrs Smith. Homework being written on the board, must try to write quickly before it is rubbed off. Bell goes, lesson ends, not too bad as lessons go, almost finished, hand aches, must have been pressing too hard on the pen again.

Lunch at last, must remember not to stand too near others in the queue; they think I bump into them on purpose; better wait till the last, hope there is still a choice left.

What lessons are there this afternoon? Check my timetable and bag, oh good PE my favourite lesson, I don't think; at least I remembered my kit. Final lesson geography, let's hope that I can finish my earthquake warning poster. Mr Peters says the idea is excellent, pity about the printing though, have to think of another way, still find typing on the keyboard difficult, could use one finger. Great, finished and looks good, no accidents or spillages.

Chantelle's experience of transition to high school

My name is Chantelle, and I recently started at secondary school. The teachers from both my primary and secondary school provide lots of help. I went with my class and friends to visit the school in the summer term. The teachers from both schools met with my parents and me.

It is very different at my new school. It is very big, and, at first, even with my map and help from my friends, I still could not find my way around all the rooms. We have to move from one class to another for most of the lessons. The things that really worry me are all the new rules, all the things I have to bring with me each day, the amount of work we have and how quickly we move on to new things in each subject. I also don't feel that other people in the class or every teacher understands my problems.

I suppose I should tell you a little about my difficulties, but that is really hard. Perhaps the first thing to say is that I often get words in a muddle; this is when I am talking and writing. Sometimes I know that it is the wrong word but can't remember the right one. Sometimes I don't know that I have said the wrong word until my friends laugh. I also find it difficult to put what I want to say in the right order when I am talking and writing. Now that I am at the high school there seem to be a lot of new words to learn in each subject. I also have to learn to spell them, which is very difficult when you can't remember the word in the first place. Reading and writing and of course talking are not my best skills, but there they are in every lesson so I will just have to get on with it. I do find it hard to listen and understand and take notes, but the teachers are very good, and I use a dictaphone, which helps me to go over them later with my Mum. Listening and talking when we work in small groups is also hard, but we have a special way of working, and so I get better every day. One of the best things is that I now have my own laptop which I take everywhere with me. It has special programmes to help me, and I am also much quicker with my work.

Michael

I was diagnosed with Tourette's syndrome when I was 12 years old. Before that time people thought that I was naughty, rude and hyperactive. Many people still do. My tics included facial twitches, shouting out and making funny noises. Sometimes the tics are really bad, and then my legs jump, my head moves from side to side, my arms swing out and I also shout out. This is very hard for me, and I try to control the movements by tensing my body or concentrating hard on doing different things, but this does not always work. Sometimes I can hold the tics back, but then it is like I am bursting; the tics explode and go on for a longer time than usual, and then I disrupt the class. My school and teachers have been very helpful, but I have been bullied. I am sometimes very angry and frustrated and hit out at other students and teachers. Sometimes I also feel very down in the dumps. I have only a few close friends, which I can understand as my behaviour embarrasses me, so it must embarrass others.

Appendix 4.1 A framework for learning through geographical enquiry

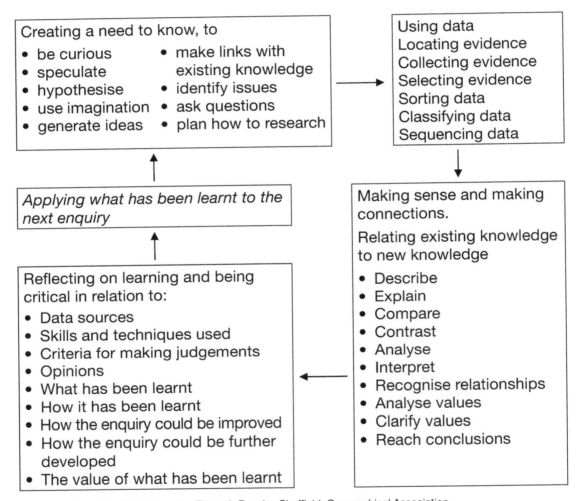

Creating a need to know, to
- be curious
- speculate
- hypothesise
- use imagination
- generate ideas
- make links with existing knowledge
- identify issues
- ask questions
- plan how to research

Using data
Locating evidence
Collecting evidence
Selecting evidence
Sorting data
Classifying data
Sequencing data

Applying what has been learnt to the next enquiry

Making sense and making connections.

Relating existing knowledge to new knowledge
- Describe
- Explain
- Compare
- Contrast
- Analyse
- Interpret
- Recognise relationships
- Analyse values
- Clarify values
- Reach conclusions

Reflecting on learning and being critical in relation to:
- Data sources
- Skills and techniques used
- Criteria for making judgements
- Opinions
- What has been learnt
- How it has been learnt
- How the enquiry could be improved
- How the enquiry could be further developed
- The value of what has been learnt

Source: Roberts, M. (2003) *Learning Through Enquiry*. Sheffield: Geographical Association.

Appendix 4.2 Risk assessment for geography field trip

Year _____ **Class** _____ **Number of pupils** _____

Date:	Destination:
Departure time:	Tel:
Return time:	Transport:
Group leader:	Tel:
Other adults:	
First aiders :	

Main activities and outline schedule:

Special considerations:

Pupil: Equipment/medicine/facilities/support

Risk levels: L = low (unlikely to occur)   M = medium (could occur)   H = could occur if adequate precautions are not taken

Activity	Risk level	Measures to reduce risk	Action to be taken if hazard occurs

Appendix 4.3 Environmental quality survey

Environmental quality survey of _____

Highlight the words that make the best description and choose a score.

Positive	4	3	2	1	Negative
Attractive, pretty, smart					Ugly, dirty, falling apart
Unpolluted, clean, healthy, pleasant smelling					Polluted, dirty, smelly, unpleasant
Colourful, vibrant, eye-catching, picturesque					Drab, dull, uninspiring
Interesting, stimulating, varied					Boring, depressing, featureless
Quiet, peaceful, restful					Noisy, confusing, unsettling

Now use the describing words (adjectives) to write your account.

The place we visited is called . . . . and
the map reference is . . . . Nearby places include
. . . . and . . . .
I enjoyed/didn't enjoy visiting . . . .
because it was . . . .
. . . . and . . . .

The best thing about it was . . .

The worst thing about it was . . .

It would be a better place to visit if . . .

Name: Date:

Assisted by

Appendix 5.1 Learning activities for assessment

Oral evidence	Written evidence	Graphic evidence	Products
Questioning	Questionnaires	Diagrams	Models
Listening	Diaries	Sketches	Artefacts
Discussing	Reports	Posters	Games
Presentations	Essays	Drawings	Videos
Interviews	Notes	Graphs	Photographs
Debates	Stories	Printouts	Recordings
Audio recording	Newspaper articles	Overlays	
Video recording	Scripts	Maps	
Role play	Short answers to	Annotated	
Simulation	questions	photographs	
	Lists	Storyboards	
	Poems	Animations	
	Descriptions	Geographical	
	PowerPoint	information	
	presentations	systems	

Consider

- Which of these are produced frequently in your classroom?
- Could you broaden the range of activities used?
- Can you add to this list?

Source: Capel, S., Leask, M., and Turner, T. (1995) *Learning to Teach in the Secondary School: A Companion to School Experience*. London: Routledge.

Appendix 5.2 How do I write better descriptions?

	How do I do it?	What do I write?
Extremes	Use place names to identify where something is. Use extremes or opposites to tell us about something.	Newcastle Under Lyme . . . Staffordshire . . . Large/small . . . wet/dry . . . Busy/quiet . . . . full/empty . . .
Different types	Recognise that there are variations or categories or different types in what you are studying.	High growth/steady growth/no growth . . . The tundra/the tropical rainforest/the desert The elderly people/ the young people with families/the teenagers/the disabled
Comparisons	Use numbers to compare features.	Twice as many people . . . half the number of visitors . . . a third less money . . . Calculate the average, the range, percentages from data.
Ratios and patterns	Spot different types; use numbers and group these together to tell us about the whole place. Try to find a pattern or relationship.	As the temperature increases the rainfall decreases . . . The further away from the town centre you go the fewer big shops there are . . .

Appendix 5.3 How do I write better explanations?

How do I do it?		What do I write?
Cause ○ Effect ○	Explain that one feature is caused by another.	A one sentence answer – adding . . . due to . . . because . . . and . . . also . . . as well as . . .
Cause ○ Cause How? Effect Effect Effect	Explain how one feature is caused by another. Or explain how one feature causes another which then has a knock on effect and causes something else.	A group of sentences – cause and effect . . . this is caused by . . . and so this means that . . . this affects this by . . . the consequence is . . .
Cause ○ How? Effect ○ How?	Explain how one feature causes another and then how the knock on effect caused something else.	A paragraph – sequencing . . . this is the result of . . . the consequence of this is . . . this means that . . . next . . . then . . . meanwhile . . . finally . . . after
○ Causes ○ Effect ⬭ Effect ⬭	Explain how two separate causes *work together* to create one feature, which then causes something else.	Two paragraphs, one about each cause – emphasising . . . the combined effect is . . . this leads to . . . these two things then create . . . this happens when . . . whereas . . . above all . . . significantly

Appendix 5.4 Writing better judgements in geography

Stage Judgement	Description
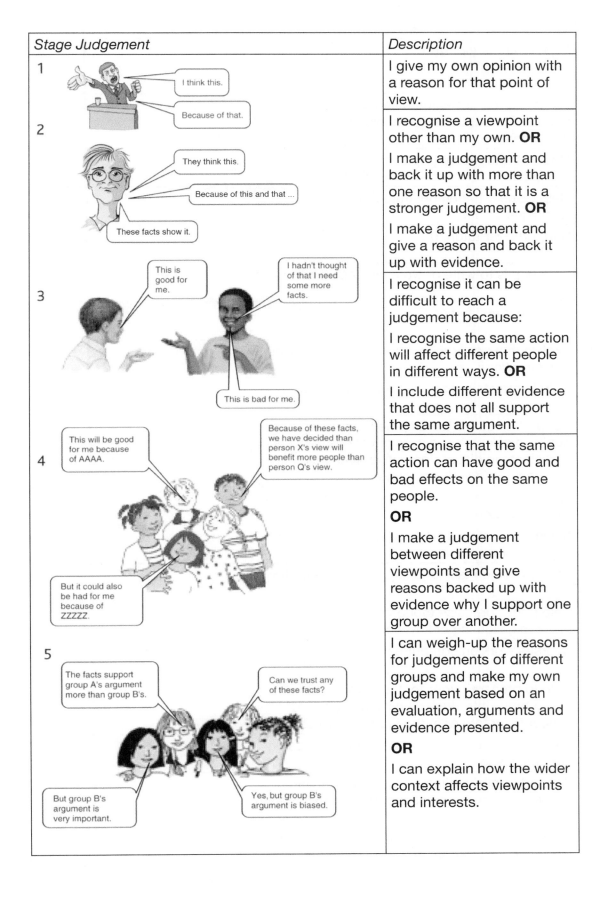	**1** I give my own opinion with a reason for that point of view.
	2 I recognise a viewpoint other than my own. **OR** I make a judgement and back it up with more than one reason so that it is a stronger judgement. **OR** I make a judgement and give a reason and back it up with evidence.
	3 I recognise it can be difficult to reach a judgement because: I recognise the same action will affect different people in different ways. **OR** I include different evidence that does not all support the same argument.
	4 I recognise that the same action can have good and bad effects on the same people. **OR** I make a judgement between different viewpoints and give reasons backed up with evidence why I support one group over another.
	5 I can weigh-up the reasons for judgements of different groups and make my own judgement based on an evaluation, arguments and evidence presented. **OR** I can explain how the wider context affects viewpoints and interests.

Appendix 6.1 Teaching assistant focus

Teacher: TA:	Date:
Lesson:	
Pupils to be supported:	*Details*
Preparing resources (modified activity sheets to simplify text, visual timetable, special equipment)	
Observe and record pupil achievements and progress	
Help teacher to demonstrate or model technique	
Support peer tutoring	
Help pupils to write or word-process work	
Check understanding	
Explain, reword or repeat concepts, objectives or instructions	
Help to maintain discipline/manage behaviour	
Support language development and encourage pupils to ask or respond to questions, participate in discussion (rehearsing their feedback, using talking frames)	
Work with a small group or an individual pupil on a differentiated task	
Manage the room whilst the teacher works with a SEN group	
Keep pupils on task and support concentration	
Help pupils organise thoughts/answers (mind maps, highlighting of key words)	
Oversee work on computers	
Help to prepare pupils for the introduction of a new topic: introduce new vocabulary and concepts (before or at the start of the lesson)	
Provide extra practice/explanation in a session after the lesson	
Support pupils' writing by providing subject-specific key words and supporting correct spelling	
Assist pupils with practical work/outdoor work, e.g. using a compass	

Appendix 6.2 Water cycle cards

The hot air containing this water vapour rises and cools	These clouds are moved towards the land by winds
The clouds rise over the land and this causes rain	This water later evaporates into the atmosphere by transpiration
Underground water slowly moves towards the oceans or appears on the surface again as a spring	The water in the oceans is again evaporated so the cycle begins once more
The rivers run downslope to the oceans	As it cools the water vapour condenses to form clouds
The sun's energy heats any water surface, e.g. oceans, and causes the water to evaporate	As the rain falls towards the earth's surface, some of it is intercepted (caught) by vegetation
The rain that falls onto the ground either soaks into the soil and rocks or flows over the surface	Water on the surface runs downhill as surface run-off and eventually flows as rivers

References and further information

Assessment Reform Group (2002). *Assessment for learning: 10 principles*. http://www.hkeaa.edu.hk/DocLibrary/SBA/HKDSE/Eng_DVD/doc/Afl_principles.pdf.

Barrett, P., et al. (2015). The impact of classroom design on pupils' learning: Final results of a holistic, multi-level analysis. *Building and Environment*, 89, July, 118–133.

Blatchford, P., Bassett, P., Brown, P., Martin, C., Russell, A., and Webster, R. (2009). *The Deployment and Impact of Support Staff Project (DISS): Research Brief (DCSF-RB148)*. London: Department for Children, Schools and Families.

Capel, S., Leask, M., and Turner, T. (1995). *Learning to teach in the secondary school: A companion to school experience*. London: Routledge.

CEA (2017). *Assess and progress*. http://ccea.org.uk/curriculum/assess_progress/types_assessment.

Coe, R., Aloisi, C., Higgins, S., and Elliot-Major, L. (2014). *What makes great teaching? Review of the underpinning research*. Durham: Centre for Evaluation and Monitoring, Durham University, The Sutton Trust.

DCSF (2008). *The assessment for learning strategy*. Nottingham: DCSF Publications. www.teachernet.gov.uk/publications.

DfE (2011). *Teachers' standards*. London: DfE Publications.

DfE (2013). *The National Curriculum in England framework document*. London: DfE Publications.

DfE (2014a). *Performance – P Scale – attainment targets for pupils with special educational needs*. London: DfE Publications.

DfE (2014b). *Statutory guidance National Curriculum in England framework for key stages 1 to 4*. London: DfE Publications.

DfE (2015). *Special educational needs and disability code of practice: 0 to 25 years*. London: DfE Publications.

DfE (2016). *School workforce in England*. London: DfE National Statistics. www.gov.uk/government/statistics/school-workforce-in-england-november-2016.

The Equality Act 2010. http://www.legislation.gov.uk/ukpga/2010/15/contents.

Haney, W. (1991). 'We must take care: Fitting assessment to functions', in V. Perrone (ed.), *Expanding student assessment*. Alexandria, VA: Association for Supervision and Curriculum Development, pp. 142–163.

Lambert, D. (2001). *Understanding assessment: Purposes, perceptions, practice*. London: Routledge.

ONS (2017). *Population and migration*. London: Office for National Statistics. www.ons.gov.uk/peoplepopulationandcommunity/populationandmigration.

Osterman, K.E. (2000). Students' need for belonging in the school community. *Review of Educational Research*, 70, 323–367.

Roberts, M. (2003). *Learning through enquiry*. Sheffield: Geographical Association.

Spendlove, D. (2009). *Putting assessment for learning into practice*. London: Continuum.

Training and Development Agency for Schools (2009). *Including students with SEN and/or disabilities in secondary geography.* http://dera.ioe.ac.uk/13793/1/geography.pdf.

UNISON, NAHT, NET and Maximising Teaching Assistants (2016). *Professional standards for teaching assistants*. Haywards Heath: NAHT Publications.

Webster, R., Russell, A., and Blatchford, P. (2016). *Maximising the impact of teaching assistants: Guidance for school leaders and teachers*, 2nd ed. Abingdon, Oxon: Routledge.

The Routledge/Nasen series of books entitled 'Supporting Children with . . . ' provides up-to-date and 'at-a-glance' guidance on teaching pupils with a wide range of special educational Needs and/or disabilities. Titles include: *Down Syndrome, Speech and Language Difficulties, Sensory Impairment, Medical Conditions, Autistic Spectrum Disorders, Dyslexia, Dyspraxia, Cerebral Palsy, Behaviour Issues.*

https://www.routledge.com/products/search?keywords=Supporting+children+with

Index